Henry Wood

The Political Economy of Natural Law

Henry Wood

The Political Economy of Natural Law

ISBN/EAN: 9783744644976

Printed in Europe, USA, Canada, Australia, Japan

Cover: Foto ©berggeist007 / pixelio.de

More available books at **www.hansebooks.com**

THE
POLITICAL ECONOMY
OF
NATURAL LAW

BY
HENRY WOOD
AUTHOR OF "IDEAL SUGGESTION THROUGH MENTAL PHOTOGRAPHY"
"GOD'S IMAGE IN MAN" "EDWARD BURTON" ETC.

The whole world around us, and the whole world within us, are ruled by Law.
THE DUKE OF ARGYLL

BOSTON
LEE AND SHEPARD PUBLISHERS
10 MILK STREET
1894

COPYRIGHT, 1894, BY HENRY WOOD

All Rights Reserved

THE POLITICAL ECONOMY OF NATURAL LAW

ELECTROTYPING BY C. J. PETERS & SON

PRESSWORK BY ROCKWELL & CHURCHILL

PREFACE.

THIS is no attempt to make people content with things as they are, but to turn the search for improvement in a promising direction. Unrest and agitation are vastly better than stagnation, but to bring the best results they must be wisely practical and in harmony with Law.

The general purpose of this volume is the outlining of a political economy which is natural and practical, rather than artificial and theoretical. While independent of professional methods, it aims to be usefully suggestive to the popular mind. As a treatise, it is not scholastic, statistical, or historic, but rather an earnest search for inherent laws and principles.

In 1887 the author issued a small book entitled "Natural Law in the Business World," which was well received and passed through several editions. The present volume is substantially a new work, although a portion of the original matter has been retained, somewhat changed in form. If it contains any larger measure of truth, the writer will congratulate himself upon any seeming inconsistency.

The different factors of society need to be drawn together and not rent more widely apart. Negative conditions exist; but they will not be improved by stimulating their realism, or by the assumption that they are inherent. Idealism is as wholesome in sociology as elsewhere. True sympathy for prevailing ills does not express itself in a morbid pessimism, but in pointing out the road to improvement and in inspiring hope and courage.

Conventional political economy, as professionally formulated, lacks a practical element which renders it of little utility in actual experience. Not being fitted into the nature and constitution of man, it is largely a mass of fine-spun intellectual abstraction. If the absorption of ponderous tomes of scholastic political economy does not add to one's equipment for the practical business of life, it is not easy to discover its usefulness.

The "cause of labor" has been injured by crowding under its banner many fallacies, and even more by the assumption that its interest is naturally antagonistic to that of other social elements. Society is a complex organism, or Greater Unit, and "when one member suffers, all suffer." The mischievous doctrine of a necessary diversity is largely responsible for prevailing frictions and antagonisms. The fault is not with the "social system," but with abuses which are the fruitage of moral delinquency in personal character. Labor and capital, when deeply defined, melt into each other.

The "labor problem" will never be solved by mere sentimental and professional treatment. The laborer often suffers more from the mistaken action of his professed champions than from the natural ills of his condition, and this will continue so long as he is led into a moral and economic antagonism. A deep and diligent search for causes and remedies should take the place of a mere superficial rehearsal of woes. Not only the human constitution, but the world in general, would have to be made over before the chimerical plans of professional "labor reformers" could be made operative. Artifice can never be substituted for evolution and Natural Law.

The writer will yield to no one in the intensity of his desire to promote, not only the public weal, but the interest of labor in its completeness. In whatever way superficial critics may construe detached statements of this book, the fact will remain that its deepest intent and animus is the true welfare of the workingman.

The recognition of the universality of Law is the greatest achievement and inspiration of modern times, and it is no less regnant in social economics than in physical science. Circumstances and conditions change, but the orderly sequences of Natural Law continue uniform. All improvement must come through a better interpretation of and conformity to its immutable lines.

CONTENTS.

		PAGE
I.	GENERAL PRINCIPLES	11
II.	SUPPLY AND DEMAND	23
III.	THE LAW OF COMPETITION	33
IV.	THE LAW OF CO-OPERATION	41
V.	LABOR AND PRODUCTION	49
VI.	COMBINATIONS OF CAPITAL	59
VII.	COMBINATIONS OF LABOR	73
VIII.	EMPLOYERS AND PROFIT SHARING	97
IX.	EMPLOYEES: THEIR OBLIGATIONS AND PRIVILEGES	107
X.	GOVERNMENTAL ARBITRATION	115
XI.	ECONOMIC LEGISLATION AND ITS PROPER LIMITS	121
XII.	DEPENDENCE AND POVERTY	133
XIII.	SOCIALISM AS A POLITICAL SYSTEM	143
XIV.	CAN CAPITAL AND LABOR BE HARMONIZED?	157
XV.	WEALTH AND ITS UNEQUAL DISTRIBUTION	169
XVI.	THE LAW OF CENTRALIZATION	185
XVII.	ACTION AND REACTION, OR "BOOMS" AND PANICS	195
XVIII.	MONEY AND COINAGE	209
XIX.	TARIFFS AND PROTECTION	225
XX.	THE MODERN CORPORATION	237
XXI.	THE ABUSES OF CORPORATE MANAGEMENT	245
XXII.	THE EVOLUTION OF THE RAILROAD	255
XXIII.	INDUSTRIAL EDUCATION	271
XXIV.	NATURAL LAW AND IDEALISM	283
	ANALYTICAL INDEX	297

GENERAL PRINCIPLES.

*"Mark what unvaried laws preserve each state,
Laws wise as nature, and as fixed as fate."*
                                                POPE.

"There is a higher law than the constitution."
                                                WILLIAM H. SEWARD.

*"All are but parts of one stupendous whole,
Whose body nature is, and God the soul."*
                                                POPE.

"When we pass from the phenomena of Matter to the phenomena of Mind, we do not pass from under the Reign of Law."
                                                THE DUKE OF ARGYLL.

THE POLITICAL ECONOMY OF NATURAL LAW.

I.

GENERAL PRINCIPLES.

BEFORE entering upon any systematic study of the inherent economic laws which permeate and shape the business world as it is at present constituted, it is well to suggest that many existing limitations at some future period may be outgrown. Natural Law is never suspended or repealed by any force which can be exerted upon the same plane; but it is axiomatic that a higher law may overcome a lower one. When we lift a weight, gravitation is not suspended, but its force is overcome by the superior law of the human will. Tree-life causes the sap to ascend, not by repealing gravity, but by surmounting it. The predominant motive of social economy, on the present plane of human development, is self-interest; but this does not always amount to selfishness, nor does it imply that individual interests are necessarily antagonistic to each other. Normal self-interest is not only honest, but entirely compatible with philanthropy. But when, in the hoped-for golden period of the future, humanity comes into a general recognition of the higher law of unselfishness, this superior force will reach down and overcome many laws that are inherent and unrepealable on their own plane. Such an advanced condition of society is to be earnestly labored for; but any present study of business tendencies must be made in the light of existing conditions and developments. Nationalists and communists, even though well-intentioned — as the great majority undoubtedly are — will never be able to galvanize unselfishness upon humanity from the outside, through governmental legislation or communistic social framework. It will only be unfolded as the natural outward expression of higher internal character.

Natural Law, as it is considered in this work, embraces in its scope the forces and tendencies which are at present operative. To hasten the evolution of higher social and economic conditions, a beginning must be made among the existent underlying antecedents which will produce them. Any inversion of this natural order will retard the coming ideal. To spend our time and energy on the outside, is only to whiten the "sepulchre." Higher attainments in any department are helped forward by the faithful use of those already actualized. When the grand reign of unselfishness is finally ushered in, it will come as an evolutionary growth, "without observation." It will be just as "natural," in its due course, as any of the lower accomplishments which preceded. Forces now operative will never be repealed in their own province, but gradually outgrown. The hope of the future lies entirely in the expansion and upliftment of character. When altruism and brotherhood are kindled in the human soul, they *will* find outward manifestation, and nothing can prevent it. All growth is from within, outward, for such is the eternal order, and no human power can reverse it. The unnatural cannot be made natural, or grapes gathered from thistles. The most ideal and perfect legislation that it is possible to conceive is powerless to raise men from the plane of self-interest. Lifting force comes from internal education and evolution.

The present "social system" — bearing in mind that its abuses are no real part of it — is the only one that will serve humanity in its present stage of development. As well fit an artificial shell to the back of a tortoise, as to frame any new external order to suit present ethical conditions and necessities. There are many such artificial shells proposed, each of which is warranted to fit — in fact to be a universal panacea — for existing ills. Among them are, land in common, governmental transportation, an income tax, limited fortunes, unlimited silver, gold monometallism, unlimited "greenbacks," a high tariff, a low tariff, free trade — all these and many more. Without any argument at present as to the merits or demerits of these proposed measures, the point is only made in this connection that it is beyond their power and range to remedy existing economic ills. If ever the time arrives when true socialism pure and simple is practical, as a form of government, neither it nor any other external system will be needed. At that high evolutionary stage every man can and may be a law unto himself. Non-resistance and unselfishness will then comprise the brief but unwritten code of humanity. At present, any new or forced artificial social framework would rather retard than aid a natural growth towards more ideal conditions.

Economic evils, now so prominent and universal, are not the outcome of the present "social system," but of the abuses which fasten themselves to it, consequent upon general moral delinquency. They are not a real part of it, but are like barnacles on the bottom of a ship. Human pride is reluctant to look within for deficiencies, but will roam to the ends of the earth to locate them outside. There is no social system, or any other human institution, so perfect, that abuses do not creep in. Stealing and cheating are abuses. They are not a normal but an abnormal part of the present order. These reflections are pertinent because sentimental theorists make our social system the scapegoat for almost every overt violation of the Decalogue.

Every genuine has its counterfeit, and every positive its negative. The present order, in its purity, is the only one for existing conditions, because it is their natural index and outcome. It fits what is back of it as the photograph represents the negative. The outer must correspond with the inner, else law and sequence would be at fault, and the chain which binds cause and effect be severed.

In political economy, as elsewhere, an intelligent study of phenomena is only possible in the light of its unseen though ever potent laws and causation. The most useful knowledge that is attainable in any realm embraces primarily the comprehension of its underlying relations and chains of sequence. The scientific standpoint from which to view human manifestations takes in, not merely present activities, but those which reach backward and forward. Phenomena are the exact fruitage of antecedents. Science formerly made but slow progress, because its attention was fixed upon superficial manifestations, while hidden beneath them were the universal and immutable forces of law. The effort has always been made to " patch up " from the outside, whereas real growth takes place in layers from the centre outward.

The phenomena of electricity have been before the eyes of the world for all the past centuries, but until recently there was little systematic study of its laws. Now that these are beginning to be grasped, it ceases to be mere uninterpreted manifestation, and becomes a tamed and beneficent agent of utility. The world has been almost surprised to find that Natural Law can invariably be relied upon. In the whole illimitable cosmos, material and immaterial, there is nothing capricious or uncertain. At first glance, there is much that seems to happen; but it may be safely assumed, that no event ever took place without an endless chain of causation leading up to it, link by link.

The scope of orderly law being unlimited, it manifestly

includes every side and phase of social economics. In the economic domain, statistics, tariffs, coinage, currency, capital, and labor have received abundant study; but all these are only the multiform visible expressions of the working of natural law. Either of them when considered by itself, outside of its larger unitary relations, becomes disproportionate and misleading. Events are unimportant except as their significance is interpreted. Statistics are only finger-boards to show the way to law-*fulness*. Their meaning and relation is the real problem pressing for solution. On the troubled surface of the sea of finance there is a confused array of facts, events, ups and downs, sentiments, and opinions, which are well-nigh valueless so long as they lack orderly translation.

If Natural Law in its immutable tendencies be reliable, and also serviceable when intelligently comprehended, it is important that its hidden leadings be searched for and discovered. But to successfully accomplish this, we must divest ourselves of all prejudice, and seek the truth for its own inherent value. Its deep lines can never be bent or distorted, but owing to preconceived theories numerous subjective illusions and inversions are possible. The desire to find a certain opinion true, often clouds the reality. To truly learn, it is necessary to unlearn. The vital truth is always beneficent, even if at first sight it have an unwelcome, or possibly an adverse aspect. To find the "whys and wherefores" of any fact is a long step towards divining its place and use. Take the law of competition. Viewed superficially — especially when applied to labor — it has hard and repulsive aspects. Shall we then deny the existence of such a law, and denounce all competitive effort as unmitigated selfishness, or not rather look deeper to see if correct interpretation will reveal utility and even beneficence? Is there constructive competition as well as that which is destructive? May it not be its abuse which is adverse, and will not a more discerning view show that it supplements co-operation? Is there not healthful competition as well as that which is unhealthful? Can there not be competitive giving, being, and doing, as well as getting and monopolizing? It is far wiser to rightly adjust any universal principle than to deny its place, or, perhaps, hastily conclude that it is only "cruel." Competition between two market-men may help to feed a whole needy neighborhood. Every thing has its place in the general unitary Whole, and when its true relations are disclosed its seemingly adverse features become neutralized or even transformed. A perfect sphere has roundness and smoothness, but its detached fragments are each irregular and jagged. A fact or principle viewed out of its logical environment does not show its truth. Any intelligent synthetic method is far too rare. To analyze, dissect, and sever, often snaps the ties of relationship and leads to unprofitable dogmatism.

Natural Law, as applied to the domain of Political Economy, is defined by Webster as "a rule of conduct arising out of the natural relations of human beings, established by the Creator, and existing prior to any positive precept." Natural Law in the economic realm is not different from that which runs through physics, morals, mechanics, and science. It is but one of the many subdivisions of Universal Natural Law, or the grand Unity of Truth. In other words, the principles which reign in the department of political economy are not artificially fenced off in a field by themselves, but they have a most intimate connection with all the other subdivisions of orderly facts. There is also a corresponding kinship in error. With false premises and a colored medium, not only one truth is subjectively transformed, but all its relations are also distorted and colored to correspond. In this way systems of negation are built up; for with one error for a basis, a whole series must be evolved to harmonize with it.

GENERAL PRINCIPLES.

Natural Law is everywhere. Its lines as they permeate the business world may not be so easily traceable as in material science, but the evidence of their existence and rule is no less positive and unquestionable. But their relations are more complex. They are so interlaced and mingled with human or legislative law on the one hand, and a purely mental and moral economy on the other, that any study of one is impossible, except in its connection. They shade into each other so perfectly that no line of demarcation is visible.

The general perception of the uniform and universal reign of law has grown with the growth of knowledge, and at the present time the highest aim of science is its fuller discovery and interpretation. Natural Law is but another name for the methods of the Creator; and that being admitted, it is evident that all just and wholesome human enactment must be founded upon it. That this true foundation is more generally recognized and built upon at the present time than in any past age, is obvious; and this is especially true where constitutional and democratic forms of government prevail. Human law is the will of society in an effort to interpret natural method; and although it may put limits on individual will, it is yet indispensable to human welfare. There has been a steady improvement in legislation and government, in proportion as Natural Law has been understood. Step by step the patriarchal, tribal, and monarchal forms of government have played their part, and led up to the modern republic, which is the most wholesome framework of society yet evolved. Further improvement will follow in proportion as the lines of Natural Law shall be wrought into the warp and woof of the social fabric.

The key to progress and approximate perfection in every department, whether physical, mental, moral, or even spiritual, is conformity to law. Take a few illustrations: A thorough observance of mental and physical hygienic law tends directly to healthful and normal individual development. A greater or less transgression brings a proportionate penalty. The penalty must be paid whether the violation be knowingly or ignorantly committed. A headache and nervous depression are very certain to follow a prolonged drunken revelry, but no more so than are panic and business stagnation to come after an era of wild speculation. That physical disease, the effect of which is to gradually thin the blood toward a watery condition, when it continues unchecked, is no less certain in its logical results than will be the degradation of our monetary system to a silver or greenback basis, if at any time a process of dilution indefinitely continues. Legislation may for a while prevent the full assertion of law, but it is nevertheless an active, living force, unceasingly pressing in the direction of its natural fulfilment. A stream may be dammed on its way to the ocean, but the final tide-level of its waters is not a matter of question. It would be as reasonable to expect to increase the efficiency of one blade of a pair of shears by the mutilation of its companion, as to benefit either capital or labor by an antagonistic policy toward the other. Illustrations might be multiplied.

Some think it practicable to transgress natural principles with impunity, so long as they avoid the open violation and penalty of human legislation; forgetting that the penalty of the former is the inevitable sequence of the transgression. One may try to persuade himself that even eternal principles are elastic and subject to exceptions, for the reason that they sometimes seem to fail to assert their rule. But if they do not vindicate themselves speedily, we may be sure that they are always pressing in that direction, and will never be satisfied till the end is reached. We confine water in a tube, but its tendency to seek a level continues, and no human power can divest it of this inclination. Natural Law is a living force, persistent, reliable, always in its place and pressing to do its work. It is this invariableness which enables us to use it, and make it serviceable. While, therefore, it is true that we are always under its sovereignty, it is no less a fact that when we comply with its conditions, it becomes our most valuable and indispensable co-worker. Its powerful aid, like that of steam or electricity, is always in waiting, only we must not dictate its methods of operation. We make mistakes, and our lines of action are often inharmonious and contrary, while the operations of Natural Law are consistent and harmonious. Its different factors may modify, or counteract, but never oppose each other, for truth cannot be in opposition to truth. Its only warfare is with error, and its complete victory is simply a question of time.

SUPPLY AND DEMAND.

"*Every natural force which we call a law is itself invisible — the idea of it in the mind arising by way of necessary inference out of an observed order of facts.*"
THE DUKE OF ARGYLL.

"*All are needed by each one ;
Nothing is fair or good alone.*"
EMERSON.

"*Extremes in nature equal ends produce ;
In man they join to some mysterious use.*"
POPE.

"*The wings of time are black and white,
Pied with morning and with night.
Mountain tall and ocean deep
Trembling balance duly keep.
In changing moon, in tidal wave,
Glows the feud of Want and Have.*"
EMERSON.

II.

SUPPLY AND DEMAND.

SUPPLY is positive, and demand negative. All negatives are seeking for satisfaction and completeness in their corresponding positives. Evil is a demand for good, disorder for order, and darkness for light. Ugliness seeks beauty; weakness, strength; and hunger, food. All positives are waiting to bestow themselves. These two principles never rest easily until united. Each will wander to the end of the earth to find compensation in its counterpart.

The law of supply and demand is perhaps the most general and fundamental of all the brotherhood of natural laws, and we have direct relations with it at all times and under all circumstances. It lies at the foundation of all modern commerce, civilization, invention, and science. It has been the main-spring in every transaction, trade, and exchange, back to the time when man existed under the most primitive conditions. It was the basis of the first exchanges of flint arrow-heads and skins among savage and barbarous tribes, as it also is of all the multiform currents and counter-currents of modern economic life. Its force cannot be measured. Its pressure impels mankind to work its behests, in gathering, transporting, and exchanging the products of the globe, in order that these two principles may meet and find satisfaction. Men will penetrate to the heart of tropical Africa, or the frigid regions of the Arctic zone; they will dive to the bottom of the sea, or delve in the bowels of the earth, to bring forth all the complex materials of supply, in order to meet the grand aggregate of universal demand. No enterprise is too venturesome, no effort too daring.

Supply and demand are like the halves of a sphere, neither being complete without the other, and each waiting for the other, as necessary to produce roundness and perfection. Throughout the whole cosmic economy each of these factors is not only incomplete without the other, but each is evidence of the existence of the other. Even in the spiritual world, universal analogy teaches that as man was created with a natural desire or demand for future existence, that this demand will be satisfied. Demand was created for supply, and supply for demand, and they have an unerring affinity for each other. A vacuum is a demand for air, and cold for heat. Man's natural constitution has many demands, and all these are easily supplied when it is in a normal condition.

Applying these principles more specifically, let us for illustration take the problem of furnishing the food supplies of a great city like London or New York. We find that just the required amount and variety are forthcoming from every quarter of the globe, and all without any system, design, or forethought. The Chinaman is gathering the tea, the Brazilian the coffee, the Dakota farmer is raising the wheat, and every other quarter and country of the globe are striving to make up the supply to fit this never ending demand. It does this as perfectly as if it were regulated by a pair of colossal balances. The element of price comes in and smooths off the inequalities, so that the two surfaces come together perfectly as though polished for the purpose. If a temporary, or even expected, surplus of any article occurs, the price drops just enough to increase the demand to the point of perfect equilibrium. If there be a temporary or foreseen future deficiency, the price rises, and the inevitable equilibrium is restored as before. It is the element of price which always determines the point at which the equilibrium is reached, and price is modified by still another element, which is competition. In the event of a tendency toward excess, competition takes place among sellers; and, on the other hand, a predominance of demand causes competition among buyers. All commercial transactions and prices, not only of material products, but of everything that has value, like rates of interest, rents, salaries, brainwork, as well as that of muscle, are so regulated. The salary of the clergyman, the fees of the lawyer, and rates of transportation, as well as wages for manual labor, are all controlled by this law. Great talent brings a high price because of its scarcity. Price is a relative quantity, and not an abstract amount. Competition among buyers may cause strawberries to bring a dollar a quart in April, and among sellers may bring them down to ten cents in June. They were relatively as cheap at the one time as the other, the price at which supply and demand became equal varying by so much in the different months.

These laws are elastic and beneficent; and they adapt themselves to all conditions in a natural and easy way, if allowed to operate without interference. Not that they will do away with all the ills of society, or give to every man employment at good wages, or always give success in business; for all such drawbacks are incidental to human fallibility and imperfection. The effect, however, of an attempt to put any forced or artificial laws in their place, is to increase tenfold the friction and difficulty. Such an effort always reacts, and is harmful to those who mistakenly hope for benefit. Let us adduce a few illustrations. Legislative interference in trying to fix rates of interest — or rather, one might say, in trying to take away the freedom of individual contract — in the different States, is now generally admitted to be worse than useless, although years ago it was regarded as necessary. The effort to substitute artificial rates for natural ones, under penalty, not only did not accomplish the purpose intended, but actually made interest dearer, by obstructing supplies, injuring confidence, and by natural reaction. When the peculiar conditions in any State make money worth really more than the maximum legal rate, the practical rate is still further enhanced, to equal the risk of the penalty which the lender incurs. Both parties also feel that they do no moral wrong by evading a statute which interferes with the first principles of personal freedom. So generally does this view of the case now prevail, that this form of legislative interference with Natural Law is practically a dead letter, although in some States the ill-advised statutes are nominally still in force. Legislative interference — except to enforce impartiality — with rates of transportation, and with passenger, telegraph, and telephone service, is in the same line, and will, in the long run, be found to produce similar results. Aside from legislative enactment (which will be considered more fully in the chapter on Economic Legislation), the most formidable attempts to force artificial prices occur in the cases of railway pools or combinations, speculative corners in food products and coal, and in labor unions. The results of these efforts are in the main unsuccessful, and in any case but temporary, and, of course, they lack the moral dignity of legislative interference. In the case of railway combinations, statistics show that in all instances where pool rates were put at a point much above that which may be regarded as normal, they were very short-lived. Such a variety of disintegrating and competitive influences come in, that even the most binding agreements to maintain artificial rates soon have to yield. In the case of speculative combinations and corners, or efforts to control market prices, it may be admitted that in a few instances they have been apparently successful, but in a vastly greater number they have not succeeded, and often have ruined their projectors. In the successful cases, where one clique of operators has succeeded in cornering the market, or in establishing artificial prices, it has only been in consequence of another clique selling for future delivery what it did not possess — in common parlance of the commercial world called "selling short," — which is in itself an abnormal condition. Any effort to artificially advance prices against natural consumption alone is rarely attempted, for to have any chance of success there must be the opposing clique of "short sellers," or those who are trying to artificially depress the market. Even with the large amount of "short selling," attempts to corner the market for food products are becoming more and more infrequent, owing to the increased rapidity of transportation, which has a strong equalizing tendency. Wherever such combinations have temporarily succeeded, the result has been brought about by peculiar conditions, and in a forcible manner, before Natural Law had time to assert itself. It was like lifting a heavy weight in spite of gravitation.

There is much popular misapprehension regarding the power which can be exerted by "combines" to change natural tendencies. We often see newspaper headlines like the following: "The West is holding back its grain;" or, "Chicago speculators are trying to force up the foreign markets;" or, "Wall Street has combined to get up a boom;" and many other similar announcements. The idea that the millions of farmers in the West, or that the thousands of operators in Chicago or Wall Street, could come to any general understanding in regard to a uniform policy is absurd. Instead of any such condition of unity ever existing, there are always two parties, known in common parlance as "bulls" and "bears," each of which is a balance to the other, like the two elements under consideration. The bears represent the principle of supply, and the bulls that of demand; and, as elsewhere, the higher or lower prices determine the point of equilibrium between them. So far from combination, not only each party, but every *individual*, is trying to excel all others in making the most correct estimate of the *natural* drift and tendency of existing conditions, and how to profit thereby.

Of forced artificial prices for labor by labor unions, we shall refer more fully in another chapter, but in passing will suggest that the uniform dominion of these principles is not suspended in the regulation of labor values, as some theorists maintain. In the long run the value of the labor of any one is determined by its relative excellency. If artificially forced up by combination or coercion it will soon react. No matter how much we might wish it otherwise, facts are in opposition. Not only that, but upon closer study we shall find that the laboring man is as much concerned in the integrity of these laws, even if he had the power to modify them, as any other part of society. As we have before noticed, the prices of brain labor are regulated by these two elements, and it would be a violation of all analogy to claim special exception in the case of muscle. He who tries to sow the seeds of discontent in the minds of laboring men by teaching such a theory, is not their true friend. He may be actuated by an honest, though misguided sympathy, but it is none the less harmful to the laborer, and tends directly to degrade his manliness and lessen his product. The sentimentalists who expect the laboring man will be benefited by force of combination — as though he were going into a combat — are mistaken. Societies of laboring men might be organized for social, intellectual, and moral purposes, and be productive of great good; but when, as at present, they are constituted for the sole purpose of forcing artificial prices, they injure not only the laborer himself, but they are harmful to business and confidence, and are detrimental to society at large. A seller of labor, as of any thing else, is dependent on demand; and demand cannot be coerced. Whenever that is attempted, it shrinks back. It is like picking a quarrel with the only friend who can help us. Supply cannot afford to repel and diminish demand. It would be a poor way to induce a horse to drink, to force his head under water.

Demand can be stimulated, courted, and increased by the adoption of such a policy as will promote peaceful conditions, and inspire confidence, for the present and the future. Wages then rise *naturally* from increased demand. Under such conditions, every employer enlarges his capacity, and as a buyer of labor has to offer higher prices to get it. The almost or quite one hundred per cent advance in average wages which has been made during the last thirty years, in spite of the immense immigration into the country, is a natural advance, and was caused by an excess of demand. If the forcing process had been continually applied during that period, the advance would have been much less marked, for the reason that the demand would have been injured. As we have already seen that supply and demand, after adjusted by price, are always equal, it follows that an injury to one is harmful to both. It may be objected that in the case of factory towns and cities, the immobility of labor would prevent in some degree the right adjustment of wages by the law of supply and demand. This may be true temporarily; but there is no other practical adjustment possible, and therefore we have no choice. However, the practical immobility is never so great, but that in the event of any forced or continued attempt to impose artificially low prices upon labor by employers, a gradual but sure process of recovery will begin at once, and not stop until the normal rate is approximated. The emigration from such a factory or town may be gradual, but it will be continuous, until the inevitable equilibrium is reached. It is no compliment to the intelligence and manliness of laboring men to assert to the contrary. The real self-interest of the employer is also a powerful factor, for the emigration would be from his most intelligent and desirable class of help.

In general, demand has grown from the cravings of primitive man for simple food and shelter, and these of the crudest character, up to the infinite and wonderfully complex variety of desire that characterizes modern civilization, and supply has paralleled its track for the entire distance. This equal progress and the enlargement of supply and demand will continue in the future, and no one can fix their limits. Until human character is evolved to that degree that unselfishness becomes the unwritten and all-prevailing law, supply and demand will always be kept equal by the regulative adjustment of *price*. May the day be hastened when the higher law *will* overcome the lower, and price be a thing of the past. But when it appears it will come voluntarily, without "observation," and for the reason that it is the outer index of transformed and illumined *character*.

THE LAW OF COMPETITION.

"*Competition is the life of trade.*"

"*Easy to match what others do,
    Perform the feat as well as they;
Hard to out-do the brave, the true,
    And find a loftier way.*"
EMERSON.

"*There are geniuses in trade, as well as in war, or the state, or letters; and the reason why this or that man is fortunate, is not to be told. It lies in the man.*"
IBID.

"*What greatness has yet appeared, is beginnings and encouragements to us in this direction.*"
IBID.

"*We must trust infinitely to the benificent necessity which shines through all laws.*"
IBID.

III.

THE LAW OF COMPETITION.

Has competition a normal place in the realm of social economics? This is a question which recently has called forth considerable discussion, and upon which opinions vary widely. In giving it an affirmative answer, we take a different and broader view than that held by some earnest and sincere philanthropists for whom we have great respect. So far as they are concerned, their practical benevolent spirit is not impaired by some abstract intellectual speculations, as to what is, and what is not, the proper framework for an ideal social system. But with all due appreciation of their altruism, the fact remains, that there is a much more numerous class of illogical people who are induced to make impractical, and even harmful applications of such speculative theories.

In forming a just estimate of any principle, it is important that a proper discrimination be made between that which is considered, *per se*, and its abuses. These, in reality, are only the negations of any system of positive good. Instead of forming any part of it, they constitute the lack of it. Normal competition is a natural law, and being deeply implanted in the human constitution, it forms an indispensable part of ethical economics. Knavery, corruption, oppression, and fraud do not belong to it. They are weak spots where there is too little of the normal present order. Average character is not yet evolved up to the level of the "system," and therefore it is the former which is at fault. When the unintelligent laborer is assured that his ills are due to the existing framework of society, he ignores the individual deficiencies of himself and others, which constitute the real source of his trouble. He is persuaded that a great institution, for which he is not responsible, is *adverse*, and hence he makes little effort at self-improvement, and imagines that he has a righteous quarrel with society in general.

To compete, as defined by Webster, is "to contend, as rivals for a prize; to strive emulously." To be competent, is " to answer all requirements; having adequate power or right; fitted; qualified; to be sufficient for." The normal use of the term does not necessarily imply unfriendliness or antagonism. There is wholesome competition in heroism, self-sacrifice, liberality, excellence of production, and in high ideals and aspirations. He who, in any position, is eminent, is competent, or, in reality, a successful competitor. To speak of competition in error, crime, or cruelty, is an abuse, or negative application of the principle. Some sentimental writers have rated it as the antithesis of co-operation. It is rather a stimulating and necessary element of co-operation, for there is competition among the most earnest co-operators. Who will co-operate the most and best? Evidently the successful competitor. Not because he is unfriendly, but relatively more competent. Competition and co-operation are the two hemispheres of one globe. They each have a necessary function in the unitary system of the Whole.

The old adage that "competition is the life of trade," is well founded. In the business world, it consists either in giving a better article at the same price, or as good a one for less. He who does these things successfully carries out the principle, and proves himself competent. The incompetent falls behind in the competitive test, and his usefulness to the community of which he is a part is therefore much less. Competition between two gas companies may give a

THE LAW OF COMPETITION. 37

whole city better and cheaper light. Though the more incompetent of the two will suffer somewhat, a thousand-fold more persons receive the benefit. Take, for illustration, a dozen leading retail dry-goods houses in any large city. A stirring competition among them gives perhaps half a million of people better goods, lower prices, a greater variety, and more attentive service. It provides for the return of goods when unsatisfactory, guarantees quality, and allows exchanges. Each makes an effort to attract patronage and to secure a reputation for reliability and liberal treatment. If six of the dozen, which are the most incompetent, suffer somewhat in the race, it is for the benefit of the half million. Many of this great community are poor, and the inevitable rivalry works greatly to their advantage. It is a partial sacrifice of the few for benefit to the many. It is deplorable that competition sometimes causes seamstresses to live in garrets and make shirts at starvation prices; but it should not be forgotten that for each one of these, a hundred poor people, as a consequence, buy their shirts cheaply. Again, were most of these shirt-makers to put aside an unfounded and foolish fancy as to relative social status, they could go to domestic service, where competition among buyers always insures not only good wages, but good homes.

Perhaps the most extreme instance of successful competition may be found in that great organization known as the Standard Oil Company. By its rare combination of skill, capital, and executive ability, it has driven a hundred, more or less, competing companies out of the business of refining petroleum. These non-competents suffer — though as a rule they have sold their plants to their gigantic competitor at good prices — but, as a consequence, sixty millions of people get better and cheaper light. There are a hundred thousand *consumers* of kerosene where there is one *refiner*. Regarding the company just cited as illustrating the power of Natural Law, we are not defending, or even considering, in this connection, the morality of its various specific transactions. If that has been defective, it is not the fault of our social system, but of private delinquency and the laxity of our legal tribunals.

Competition is not limited to individuals and corporations, but its quickening impulse is felt by states and nations. Wherever it is most prevalent and intense, there the progress of science, invention, and civilization is the most rapid.

Any effort, in the business world, to excel in giving less, a poorer quality, or at a higher price — in short, to render an inferior service — is not competition. Such an effort would be only its absence, which might well be denominated *in*-competition.

The sentimental prejudice existing against this universal law is the result of a narrow view of a single element in it, as seen disconnected from its relations. It is mainly indulged in by those who are unaware that law pervades the economic realm, but are of the opinion that it is governed by sentiment and impulse.

Were it not for the ever-present stimulus of the desire to excel, we might still be travelling in clumsy wagons without springs, instead of the "limited express." We would still navigate the sea in "caravels" — if not by means of even cruder craft — unmindful of present luxury and speed, which make distant nations like next-door neighbors.

China has made little progress in art, science, and invention for thousands of years, from the almost utter dearth of emulation among its people. Barbarous tribes emerge from savagery, ignorance, and poverty, only as rapidly as competition becomes mingled with, and added to cooperation. An Indian tribe has the latter in high degree, but as the former is wanting there is little individual progress. Co-operation alone, keeps all upon the same level. Each is satisfied with the methods and attainments of his ancestors and neighbors. Competitive energy has evolved the whole fabric of modern civilization. A watch could as well run without a mainspring, as the world make progress with this universal tonic wanting. But a sickly and sentimental paternalism sees competition as a principle which is inimical to the interests of the "laboring man." Is such a man disconnected from, and an exception to, all the rest of humanity? One would so conclude, for his champions often assume that his welfare lies in direct opposition to that of the rest of society. But the fact is that competition is his best friend. It impels him toward the very industry, merit, and progress which his self-imposed leaders and guardians discourage. The desire to excel in the laboring man is the great lever to lift him higher. But for that he would forever delve on the low plane of mediocrity. It prompts and spurs him to better service and higher attainment. It uplifts, not merely by pushing from behind, but through ideals from above. Its motto is ever, Excelsior! Every one who is competent, and competing, gives the world his co-operation by raising the general average.

The term "laboring man" has been mistakenly limited in its popular significance, so as only to include those who labor in a particular way. The law of labor is both universal and beneficent; and he who strives to evade it, and does not in some way *work*, in and for the world, will suffer for its violation.

The disparagement of competition by labor unions is a costly mistake, for it virtually puts a premium upon incompetency. Individual excellence, and an ambition to rise above the dead level of other incompetents, meets with thinly disguised disapprobation. A most baseless and mischievous theory has attained wide acceptance, that, in the process of rising, one necessarily pulls another down. The exact opposite is true, for every wholesome example really forms a general ideal and stimulus. Prevailing fallacies directly lessen and deteriorate product, while the irrepealable law ever remains, that in the long run excellence alone confers value. The combined labor unions of the world cannot permanently lift wages above their natural value, neither can the united "trusts" of the world confer abnormal value that will last. Quality and demand together form the only permanent basis of value in the commercial world, and the united force of combination and legislation cannot render it otherwise.

Were it possible to do away with the law of competition, humanity would settle down to a stagnant level, and evolution be turned back.

When, in the future, mental and spiritual evolution shall have ushered in the ideal reign of unselfishness and altruism, there will still be an active competition in kindly deeds and loving ministry.

THE LAW OF CO-OPERATION.

"*In union there is strength.*"

"*United we stand, divided we fall.*"
MORRIS.

"*Men will live and communicate, and plough, and reap, and govern, as by added ethertal power, when once they are united; as in a celebrated experiment, by expiration and respiration exactly together, four persons lift a heavy man from the ground by the little finger only, and without a sense of weight. But this union must be inward, and not one of covenants, and is to be reached by a reverse of the methods they use. The union is only perfect when all the members are isolated. It must be ideal in actual individualism.*"
EMERSON.

"*A long pull, a strong pull, and a pull all together.*"

IV.

THE LAW OF CO-OPERATION.

In the preceding chapter the fact was noted that the laws of competition and co-operation supplement, but do not antagonize each other. Either, taken alone, is fragmentary and incomplete. Like supply and demand, each rounds out its counterpart. They are therefore friendly, both being indispensable in the normal Whole.

Every orderly commercial transaction includes the co-operative principle. Though one may be a buyer and the other a seller the action is concurrent. The true co-operative spirit in commerce presupposes that both parties are benefited. A farmer exchanges a load of potatoes for a coat. His need and that of the clothier are both supplied, and the trade is co-operation. Commerce always implies both competition and co-operation. It is not true that in business transactions, one is naturally a loser and the other a gainer. Both should gain. Where it is otherwise, the fault is not in commerce, law, or system, but in individual judgment or integrity. If one is cheated, there is a lack of co-operation. There is also a want of competition, for if this element be fully present it is a protection against loss, or a "poor bargain." In every normal commercial transaction both parties are co-operators and also gainers, because surrounding competition furnishes a kind of guaranty that neither shall lose. Viewed in itself, then, the present economic order, with all its inherent laws, is beneficent. Commerce, *per se*, is altruistic. All possible failure must therefore find its location in individual character.

Co-operation as usually defined is "concurrent effort, or labor; operating jointly with another." It is a universal and indispensable law in man's social constitution, and in the world of economics. "In union there is strength." Without this principle no notable human accomplishment would be possible. Every great edifice, city, railroad, or manufactory comes into existence through adaptive and general co-operation. Even the simplest human product embodies the co-operative principle. A needle bears silent witness to the existence of a great factory where there is found a variety of concurrent effort employed for a single end. A great artist would be helpless without the co-operative skill of the manufacturer of pigments, concerning the production of which he may know nothing. The principle is so complex and silent in its operations that we are largely unconscious of its ubiquity.

Co-operation lies so deeply imbedded in the human constitution that it may be called an instinct. It is therefore common to all the inferior planes of life. A swarm of bees furnishes a good illustration. Below the level of humanity each co-operative circle is naturally confined to those of the same species. The co-operation of the bee does not extend beyond bees, or, perhaps, is still further limited to its own swarm. Animal co-operation uniformly has narrow limits. When, in the ascending trend of organized life, man is reached, the principle broadens, and its normal and ideal position as a *law* begins to be realized. In proportion as animality is slowly but surely overcome, co-operation will continue to widen its scope until it becomes in action what it now is ideally — all-comprehensive.

It is undeniable that from a superficial standpoint, co-operation has a more attractive and unselfish aspect than competition. But a deeper view shows that limited co-operation usually has a basis of self-interest, if not of selfishness. As popularly defined, its application is always restricted."

We co-operate with those of *our own* union, sect, secret society, or political party. But what of all the rest of mankind? If a man does not belong to *our* union, he is a "scab;" if not to *our* party, he is a "demagogue." If we live in Colorado, Eastern people are "gold bugs," or, if in New York, those of the West are "silver lunatics." Not only co-operation, but the true co-operative spirit is held under restriction. Just in proportion that such is the case, it becomes selfish and antagonistic. Limited co-operation really means co-operation's negation, or absence, and thus it entirely fails to fulfil its ideal function. Prevailing co-operation is still that of the animal plane. Until limitations are outgrown it will be only elementary. Education comes through an adverse experience among partial negations and rudiments.

But *all* limited forms of co-operative effort are by no means to be condemned or discouraged. They are abuses only to that degree that they embody a spirit of antagonism toward the rest of society. An organized body of carpenters whose mutual aim and interest is to develop moral and intellectual fibre among its members, to provide for their social recreation and enjoyment, to succor them when in distress and to stimulate technical and mechanical skill — all these embrace the true co-operative principle, though limited in scope. Such elements are positive and their limitations are only superficial. Their spirit overflows the boundaries of their visible application. Unions that operate upon such lines are working in harmony with broad co-operative law. But when such aims are set aside, and an unfriendly spirit developed towards those of the same trade who, in the exercise of individual freedom, do *not* belong to *our* union, there is a want of true co-operation. When such a temper becomes ruling and those of other, or no unions, are abused or threatened, it is purely animal co-operation, which is organized selfishness.

A co-operation, though limited to those of the same profession, trade, or calling, which seeks to increase its efficiency in service to the greater unit — society at large — is proper and laudable. The carpenter must not narrow his view of the carpenter's interest entirely to his own trade, because his real interest is a part of, and bound up in the general interest. The larger is more important than the smaller, and the whole than any one of its parts. This truth, though so fundamental, finds but little lodgement in the popular mind. Society, instead of being a great co-operative, harmonious One, is split into warring fragments. It should be like a grand orchestra where many dissimilar instruments co-operate for the elaboration of one supreme theme. Suppose that the violins co-operate alone, what becomes of the symphony? But some theoretical socialist will suggest that the orchestral illustration is quite in the line of his own philosophy. Let us see. Could the enactment of the best musical legislation, and a proclamation of all the laws of harmony to the united orchestra, enable it to interpret one of Beethoven's immortal productions? No, there must be a preparatory individual education, and an inner inspiration, and only when these have formed the basis, is an orderly and harmonious expression possible. The music must flow forth in vibrations from within. It cannot be artificially imposed from without. All true unity comes from a drawing towards a common centre, and not from external binding or hooping. If true co-operation could be inaugurated by legislative enactment, a single day would suffice for the accomplishment of a great evolutionary age-long process.

The human body, in its normal condition, is an ideal illustration of the co-operative principle. The body is *one*. No single set of members organize a union against the others. Concurrent effort is not limited to those having a like office, but embraces the most distant and unlike. Though some appear less important, yet all are needed and honorable. Paul delineated this beautiful relation not only religiously but scientifically.

But a necessary and expected antagonism between the various members of the body-politic is everywhere assumed. Conventional literature, the daily press, common habits of thought, and general consent, all combine to create and emphasize a universal disagreement. It is everywhere taken for granted that the *interests* of different sections are inimical to each other. That of the farmer is against that of the manufacturer, that of the importer opposed to that of the exporter, and, more than all, that of the poor contrary to that of the rich. Class prejudice, which can only aggravate existing evils, is systematically stimulated. Ignorant and fanatical self-constituted leaders build up a fallacious political economy, and gain a cheap notoriety by arousing section against section and class against class. Trades, professions, unions, parties, and societies are led to believe, in all sincerity, that *their* interest is peculiar, and disconnected from the common interest. Friction is everywhere increased, prosperity blighted, and confidence destroyed.

Paul's illustration of co-operative activity among the bodily members holds equally good in the business world. The law is immutable, and seeming temporary and superficial exceptions do not in the least invalidate its deep, silent trend. They are only eddies on the bosom of a great river.

Law, complied with, brings harmony, and harmony introduces prosperity. Amid all the jarring contentions of various "interests," the unrepealable principle remains that no "member" can more than temporarily suffer or rejoice *by itself.* The rule holds good when extended to nations. Any seeming advantage to one, gained through the misfortune of another, is only superficial.

In the economic world all parties and transactions have real though invisible relations; but only in proportion as these are permeated with a genuine co-operative spirit, will harmony and prosperity prevail.

LABOR AND PRODUCTION.

" He that by the plough would thrive
Himself must either hold or drive."

" Man goeth forth unto his work and to his labor until the evening."
Ps. clv. 23.

" On bravely through the sunshine and the showers,
Time hath his work to do and we have ours."
EMERSON.

" If all the year were playing holidays,
To sport would be as tedious as to work."
KING HENRY IV.

" Hear ye not the hum
Of mighty workings ? "
KEATS.

" Human labor, through all its forms, from the sharpening of a stake to the construction of a city or an epic, is one immense illustration of the perfect compensation of the universe."
EMERSON.

V.

LABOR AND PRODUCTION.

LABOR is normal; idleness, abnormal. The physical, mental, and moral faculties of man were created for use, and it is only by their active training that they attain skill and excellence. That the active employment of the gifts and capabilities of man's nature was designed by the Creator is proved abundantly, both by analogy and experience. As all human happiness and perfection are reached by conformity to law, so a non-conformity brings misery and unhappiness. Labor is a blessing, and idleness a curse. Human powers must have occupation, else they become withered and inharmonious. As man is constituted, it were better to give him the barren and sterile soil bringing forth weeds and thistles, to be transformed by the healthful activity of his energy into blooming gardens and fruitful fields, than to supply him with all these delightful and useful objects without effort and toil on his part.

The world is full of positive possibilities, and honest labor is, therefore, the most staple of all commodities. The mistake of thinking that only manual labor is labor is a very common one, while the fact is that every power of the body and mind requires exercise; and only by this activity can they fulfil their offices. Under primitive conditions, there was a general activity of body and mind, rather than special development in any one direction. The barbarian was his own tailor, carpenter, jeweller, farmer, and common carrier; and his products were few and poor. Under modern conditions, activity is greatly subdivided, and education much more thorough. Thus we have farmers, carpenters, painters, engravers, masons, and numberless other craftsmen, each one of whom has a special technical training, and, as a result, decided superiority. Each, therefore, does not only his own particular kind of work for himself, but for all the others, because his production is far more perfect. So in the department of mental labor: the clergyman, lawyer, banker, scientist, historian, and statesman — all cultivate their powers in their several fields to a high state of efficiency; and each has his place in rounding out and completing the grand unit called society. In this consists the great superiority of the modern state, with its high degree of specialized education, over the barbarous governments and peoples of primitive conditions.

The scientist, historian, and bookkeeper are as truly laborers and producers as are they who handle a pick, plough, or loom. The popular use of the term "labor" as applied only to those who exercise muscle is erroneous. The brakeman in the employ of a railway company, by industry, energy, and ability, may rise to be its president, but he is no less a laborer than before, and as a man not necessarily any more worthy or noble.

While a normal amount of labor is in accord with law, and is necessary to healthful and harmonious development, an excess of exertion is harmful. It is also obvious to any close observer that, of the two, undue mental effort is more wearing in its results on the health and constitution than too much physical exertion. The care and responsibility incidental to mental occupations cause many to break down in health; and here again the popular idea is at fault that connects all hardship and suffering only with manual occupations. While, therefore, our sympathy goes out towards the laborer who uses a shovel for eight or ten hours in a day, we should not entirely overlook the weary bookkeeper or clerk, who often works much longer amidst unwholesome conditions and impure air. The sleep of the man who exercises muscle is sweeter, his digestion more sure, and his vigor greater than that of the average mental laborer. The idea that manual labor is in itself degrading, and to be avoided so far as possible, is the delusion of the present time.

The ideal man is he whose physical, mental, and moral powers are all cultivated and harmoniously balanced. Idleness is a violation of natural law, and its companions in transgression are improvidence, degradation, intemperance, and decay. By inexorable law and logic each positive virtue has its corresponding negative condition of vice and error.

As to the different varieties of labor, all are indispensable, the mental as well as the physical, each in its proper sphere. The steam in the locomotive is a more subtle and immaterial factor than the boiler and wheels, but no less necessary and important. So the mental worker, though in a more refined and nominally higher sphere, is only a component part of a general system, and in personality is not necessarily above his fellow laborer. The test of the excellency of a wheel in a machine is that it fills well its peculiar place and office.

Having found that work is natural, necessary, and in harmony with man's constitution, let us consider its object. In the economy of Natural Law, means are always in order to ends. Labor is the means; production the object. The finished building is as much the product of the architect as of the carpenter or mason; or, rather, it is the joint product of all. In the distinction made between mental and manual labor, it is evident that only the predominant element is referred to, for neither can be strictly pure. The simplest manual task must be accompanied by a mental process; and likewise, the scholar or scientist must do some physical labor with pen or apparatus.

Production is only a general term for food, clothing, home, education, surplus. These constitute wealth, which is only another name for accumulated labor. The wages paid for labor are rather the above-named objects, than any certain sum of money, for the value of money consists only in the products that it will command. The natural aim of the laborer is to increase the result produced by his effort. How can this be done? First, subjectively, by greater activity and through the cultivation of individual qualities which tend to success. Second, by surrounding himself with more favorable environment and conditions. It is not only in accord with Natural Law, but also with common sense, that individual energy and thorough training in a particular department are necessary for much progress in that line of effort. The question with the wage-worker should not be, how few hours or how little exertion can I possibly get along with? but rather, how much can I accomplish? He who puts forth his best efforts will soon become indispensable to his employer, and his labor will naturally increase in value, and he himself by positive development will become an employer.

Society is composed of two classes, the independent and the dependent. To which of these two classes a man will belong is, under all ordinary conditions, a matter of individual choice. The terms independent and dependent are here used in a relative and not absolute sense. There is no absolute independence, for interdependence is universal. But relatively, every one who mingles faithfulness with his labor and keeps his expenditures within his receipts, is economically independent. This is true without much regard to the amount of difference, provided the margin be on the side of thrift. It is true that the situation of a wage-laborer is sometimes subject to contingencies, but with rare and local exceptions, conscientious labor is always in demand. The higher it becomes in quality, the more scarce it grows in quantity. The high grade is never plentiful, hence demand meets it on an elevated level, both as to value and stability. The highway to independence is open, and guide-boards are up at every turn. Just here is seen one of the bad effects on the laborer of actual — not ideal — labor unions. A member, instead of depending upon individual merit and energy for maintaining or advancing his wages, relies upon the power of the union. The former is natural, the latter artificial. By this course he loses his motive for the attainment of personal superiority and natural advancement, and settles down to the dead level of the dependent elements which surround and control him.

The goal of the American laborer is the position of accumulated labor, or, in other words, that of proprietor. A continuous, even if small margin between income and expenditure in one direction, fixes the condition of independence, and, in the other, of its necessary opposite. It is not a matter of chance, but of law. In this country, even if a laborer begins in the dependent ranks, his condition is not a fixed one. The transition to the independent class is easy and plain, when the natural course of individual merit and effort is chosen. Examples on every hand prove that this is a universal experience, and not a matter of sentiment or theory. But a very small part of the wealth of this country was inherited, probably nine-tenths being the result of personal enterprise. Any short-cut route to success is uncertain, and any forced march, outside of the natural conditions of progress, or under a dictator, is generally disastrous. But the broad, direct, and solid highway of individual industry, economy, and temperance is open always. A surplus is what the daily wage-worker should be accumulating, and presently it supplements his personal force with power of another kind. For such a man to try to antagonize accumulated labor, or those who possess it, is to oppose the very principles and conditions which are his own hope and reward.

The young American wage-worker who puts forth his best efforts, and who practises what economists call abstinence, or the limiting of expenditure to less than income, has as good ground for expecting to become a capitalist as has the gardener to expect a crop from good seed deposited in rich and fertile soil. It is no less true that he who does as little as will possibly keep him in his position, and who has slight regard for the interests of his employer, has the elements in him which make it almost certain that he will be always a member of the dependent class.

In regard to means favorable to increased production by labor which are external to the laborer, two general conditions may be mentioned: first, that of increasing the efficiency of mechanical appliances and aids; and second, seeking a favorable location or propitious field for operations. As to the first, it is not long ago when labor-saving machines were looked upon as the enemy of the laboring man, and some of the most useful inventions were forcibly destroyed, and their owners persecuted. Even so recently as thirty or forty years ago, the opinion was quite prevalent in the rural districts of New England that the general advent of railroads would quite destroy the value of horses and oats. It was found later that the world needed both, and the result was just the opposite of what the farmers had expected.

When the printing press was first brought into use, it was found that with it one man could do the work of two hundred copyists, and, as a consequence, it was feared that one hundred and ninety-nine men would be thrown out of employment. But what was the result? Soon the superiority of printed over written books, together with the lower price, stimulated authorship and increased the sale and use of books a thousand-fold, and employment was given to more printers than there were copyists before. Besides this direct result, there were in addition the related occupations of paper-makers, book-binders, book-sellers, and various others, so that the final outcome was the demand for many times the number of persons who seemingly lost their occupation when the invention came into practical use. And this, as the result merely of an economic process, aside from the immense impetus given by it to learning, art, and science. It may be regarded as in accord with Natural Law that every new invention and improvement which saves manual labor, and adds to the comfort and convenience of mankind, at the same time increases and opens up new avenues of employment, so that, as in the instance just noted, it gives occupation to a greater number of operatives than were before required. This may require time, but the process is steady and sure. Every improved appliance not only increases production and adds to the varieties of occupation, but it also raises the grade of employment. The engineer who runs a locomotive has a higher quality of occupation than he who wields a pick, for the reason that it includes more of the intellectual element. The superiority is in the relative grade of production, for the man who uses the pick is not necessarily lower or less honorable as a man.

In general, with the progress of science and invention, mind has more and more asserted its supremacy over matter, and the physical exertion of the laborer has been tempered in an increasing degree with the intellectual element. An ever increasing proportion of the aggregate work of the world is of the mental variety. As man becomes better acquainted with Natural Law, he gains in his supremacy over, and command of the material elements around him, and makes them minister to his complex needs and desires. All this is to the special advantage and benefit of the manual laborer. In consequence of this, the humble cottager of to-day has more comfort and even luxury than the king in his palace could have enjoyed three hundred years ago. It is said that Queen Elizabeth wore the first pair of knit hose ever brought to England, and they were regarded as a great luxury; while now even a beggar could hardly be found without them. The introduction and use of the telegraph, telephone, and other electrical appliances afford examples of the conveniences now enjoyed by all classes. These, at the same time, open up immense fields and new avenues for human energy and employment. As before suggested, labor becomes more efficient in production by subdivision. The Jack-at-all-trades quality of production belongs to a past age; the present tendency being towards perfection of detail, by means of thorough organization and subdivision.

The law of progress is in the line of each member of society doing the particular thing which he can do best, and leaving everything else alone. This natural principle is being widely utilized, and, as a result, no past age can be compared with the present in respect to the ease, quantity, and quality of production.

Our own country, without doubt, presents a field of operation where the greatest possible production can be gained from a given amount of labor. The American youth have before them the most promising opportunities which have ever been enjoyed in any age or country. They are indebted for this, not only to the fact that they have the command of all the accumulated skill, knowledge, and experience of their predecessors, but that all their natural rights and privileges are secured to them by the beneficent care and protection of free government. They start in the race without any of the impediments that pertain to less democratic conditions. In the Old World, the fixedness of class, rank, and position, together with its systems of entail, compulsory military service, and many other influences which are artificial in their character, are dead-weights, and in opposition to the free exercise of Natural Law.

COMBINATIONS OF CAPITAL.

"*Where large management is more economical and productive than small management, we shall find large concerns or none at all. Business, even in those lines where there is partial monopoly, is carried on with too narrow a margin of profit to endure any but the most economical methods. To control the abuses without destroying the industries is a matter of the utmost difficulty.*"

A. T. HADLEY.

"*A dispassionate view of the subject will, in my opinion, convince a competent person that the general economic function of a corporation is to perform steadily, cheaply, and permanently, a service which an individual can only perform briefly, and with comparative inefficiency.*"

C. S. ASHLEY.

"*Thus is the problem of Rich and Poor to be solved. The laws of accumulation will be left free; the laws of distribution free. Individualism will continue, but the millionnaire will be but a trustee for the poor; intrusted for a season with a great part of the increased wealth of the community, but administering it for the community far better than it could or would have done for itself. The best minds will thus have reached a stage in the development of the race in which it is clearly seen that there is no mode of disposing of surplus wealth creditable to thoughtful and earnest men into whose hands it flows, save by using it year by year for the general good. This day already dawns.*"

ANDREW CARNEGIE.

VI.

COMBINATIONS OF CAPITAL.

The conditions of to-day afford opportunities and inducements for the combination of capital unparalleled by any past era, and yet, equalizing forces have grown powerful in a corresponding proportion. Corporations are by far the most common forms of capitalistic combination; but as they are considered in another chapter the present study will be given more especially to those modern commercial phenomena, popularly known as trusts and corners. The former are usually concerned in the manufacture of product, as well as its disposal, and the latter more exclusively in the manipulation of market values. Railroad consolidations, which are essentially great combinations of capital, are also noticed elsewhere.

It is not easy to consider the conventional trust or corner from a single stand-point. There is the ethical, which is somewhat distinct from the economic view. These shade into each other in indefinable degrees, but are not identical. While the working of natural economic law is mainly considered, its ethical relations are intimate and should not be overlooked.

Not only from a humanitarian but from a purely moral stand-point, all combinations which make an effort, whether or not successful, to force abnormal values, deserve condemnation and that only. The writer desires to emphasize this conviction because, while he attempts to show that they are "cabin'd, cribb'd, confined," by Natural Law, and have not a tithe of the harmful power — except to those who invest in them — that theorists have credited them with, at the same time he utterly denies any supposed defence of, or excuse for them. Let us regard the question as settled that *all* attempts to impose artificial prices for products are morally delinquent, even when technically legal as tested by legislation. The immorality is in the intention, whether it be wholly or partially successful, or an entire failure. But in any case where a trust or combination is formed for the purpose of greater economy, or cheaper production, and with the aim of selling its product at fair and not abnormal value, it has good reason for existence and is ethically sound.

There is perhaps no economic topic upon which there has been more unintelligent and superficial reasoning than upon combinations of capital. Theorists, including many clergymen, with good intentions, but ignorant of the practical laws of business, have vastly overrated the power of combined capital.

Let us try to arrive at the truth, for that cannot be unfriendly to labor. It is everywhere and always a saving force. There is nothing that more surely clouds it than a sentimental pessimism.

In a brief study of combinations of capital, as related to inherent social and economic forces, we will begin with the modern trust. Such combinations have been painted in lurid colors as dangerous monsters; but in reality, owing to natural limitations, no one but the stockholder need fear them. He is the victim of flaming prospectuses, unsound principles, and of an inordinate ambition to gain by a "short cut." The adverse experience of thousands, who have imagined that combination could override such an antique law as supply and demand, is very uniform. Many excellent financiers, so considered, have utterly over-estimated the power of combination, and realized loss through investments in trusts, instead of expected gain.

But the trust principle may be employed in accord with

Natural Law and the public interest. To increase production, cheapen processes, and improve quality by systematic consolidation is lawful. A great university is an educational trust. Instruction in specialties, scientific experiments, and thorough research are all more exhaustive than would be possible in half a dozen small disconnected colleges. Likewise, factories, even though located apart, by becoming a unit may sometimes introduce new economies so as to better serve society in serving themselves.

But with the great majority of trusts now existing, as well as many which have come to an untimely end, the above enumerated objects are secondary or incidental, while the supreme motive is to impose abnormal values upon product. How many well-laid schemes of this kind — schemes that looked so formidable as to excite apprehension — have dissolved through the working of silent but immutable forces! Their collapse, or disintegration, was purely a question of time. Their sanguine projectors, who thought it possible to raise a few square miles of the ocean above the general level, have exchanged a financial loss for an educational experience. A trust, even if inflated, may run smoothly for a short time, but presently some unlooked-for competition or monetary stringency dries out the "water" that gives it proportion, and it shrinks to its normal leanness. When the collapse comes the law of reaction brings its value down, temporarily, nearly as much below as it has been above the normal level.

It is true that during the recent era of industrial combination much profit has been realized in the *organizing* process of trusts. Suppose half a dozen factories, more or less, engaged in the same line of production, have a normal value of a million of dollars. They are consolidated into a trust and capitalized at three millions. If the original projectors are able to find confiding and visionary investors who will buy their stock at par, they have made two millions. Such a profit is immoral and in violation of Natural Law, even though it be technically "legal." Such wealth is gained, not by the *operation* of a trust, or from the general public, but before it fairly begins business. The loss is that of the investors and not of the community. The latter is protected by natural limitations. Consumers are safe because the most powerful combination can bolster up abnormal values only temporarily. Unseen and untiring forces are fighting against it. Demand falls off and competitive production is stimulated on every side. The market is soon overstocked, and values go down with such momentum that they do not stop until consumers have their compensation, with interest added. A few typical examples may be of interest. In 1888 a trust, or syndicate as it was popularly called, was formed, with headquarters in Paris, for the purpose of gaining control of the visible and future supply of copper, and for the manipulation of its market price. The operations of the great "French Copper Syndicate" are almost of dramatic interest, not only because they are thoroughly typical, but for the reason that the contest was the most gigantic one in modern commercial history. On one side was unlimited capital and eminent financial ability, while arrayed against this combination were only the silent and unseen forces of *Law*. The principles involved, being identical with those of scores of lesser trusts, a brief outline of the operations of the famous syndicate will serve for illustration. When the combination was first formed it had no ostensible purpose of obtaining world-wide control of the metal in question, but, after advancing to a certain point, found it necessary either to retreat at a decided disadvantage, or go forward with the purpose of gaining full control of the main bulk of the entire copper product. Note the conditions. The field of operations included but a single metal, and that one having but limited sources of supply. The resources of the syndicate were immense, including some of the largest banking houses of Paris and London, with ramifications extending to America and elsewhere. The names of financiers of world-wide reputation were prominent in the management. The most extravagant expectation of profits was entertained, not only by the projectors, but by the commercial world in general. It was like a gladiatorial contest between giants. The normal value of copper at the beginning was about ten and a half or eleven cents per pound. The syndicate began its purchases quietly, storing the metal for a rise. It easily secured the bulk of the visible supply, and made contracts with producers for future product at gradually advancing prices. It withdrew the metal from the open market and went on with the accumulation until the expected high prices of the future should enable it to realize great profits. It reasoned that the world *must* have copper, and would be obliged to come to them and pay their prices for it. This logic was satisfactory to the best (so supposed) financiers of the commercial world. Who should or could arise to dispute its supremacy? Not any personal or corporate opposition, but invisible forces which are as unrepealable as the tides. The syndicate serenely paid for its copper and laid it away for the good time coming. Month by month the price was advanced. By fractions it steadily moved from eleven to twelve, thirteen, fourteen, fifteen, sixteen, and seventeen cents per pound. Copper mining everywhere was greatly stimulated, and the output increased in unexpected degree. Consumption waned, not by any agreement among consumers, but because of abnormal prices. With each advance less copper was used, and wherever possible some other metal substituted. In the meantime a mountain of shining ingots was being piled up in syndicate warehouses. What should be done with it? Scores of millions were already invested, and still the process was constantly accelerating. The longer the day of retribution was postponed the more intense would it be. It came, and down went the price and the syndicate. Flood tide was followed by ebb — a very low ebb.

Natural Law punishes its offenders without the aid of courts or judges. Every principle involved in the operation and collapse of the French Syndicate is alive and unceasingly active in every lesser attempt to override inherent forces. Even were the entire wealth of a first-class nation put into the effort, the result would be the same. "Supply and demand," almost outlawed by modern theorists, still retains its pristine vigor.

The mania for trusts has been general and suicidal. The insane desire for sudden wealth blinds the vision to adverse object lessons on every hand. A "short cut" to affluence is the *ignis fatuus* of the modern business world. There are combinations in sugar, cordage, cotton-seed oil, matches, biscuit, strawboard, and numerous other products. The midsummer financial cyclone of 1893 swept through them and, generally speaking, left them shrunken and dilapidated. In some cases they may again gather a little energy, but it is evident that the "vertebral column" of the industrial trust mania is seriously fractured. Panics, like thunder-storms, purify the air.

A dry goods establishment, soap factory, or brewery cannot increase *earnings* simply by the magic of transformation into a stock company with an inflated capital. Oftener there is a retrograde, in consequence of less careful management and diminished economy. However, a most commendable exception is found when a private enterprise is merged into a stock company, *at a normal valuation*, for the purpose of a general participation of interest by employees.

There are a few important combinations having peculiar features which deserve mention. Prominent among them are the Sugar and Standard Oil trusts. The former has been nominally transformed into a single corporation, but it still retains the essential trust features. Circumstances up to the present time have made it more than ordinarily successful. While its shares have been a prominent "football" in the stock exchange, it has paid liberal dividends, and just now seems quite well established. What are its peculiarities? Sugar refining is an industry, the establishment of which requires an unusual length of time and exceptionally large capital. Every refining plant, however, that may be built in future will have to be taken in by the combination on some terms, or competed with. Just in proportion that artificial values are imposed on product, the projection of new refineries will be induced. With unlimited millions of idle capital seeking permanent investment on a net four per cent basis, enterprise will not be dormant, and the public will be increasingly protected. There are other collateral influences which have given the sugar combination some advantage over trusts in general. Among them may be noted, eminently able and conservative management and executive ability, and operation in a product of universal and increasing demand, together with a less excessive capitalization in proportion to the value of its aggregate output. Without here considering the merits or demerits of the present tariff, it is evident that forced artificial values of much magnitude would cause regulative importations. The prediction is ventured that in the not distant future, competition in sugar-refining will so increase that dividends will at length be reduced to the basis of a fair return upon a *normal valuation* of the combined plants. Natural Law will never rest its forces until all inflation has been atoned for and rectified.

What are the peculiar elements which not only have preserved the Standard Oil Company from disintegration, but given it unprecedented success? Before attempting their enumeration, it may be proper to suggest, that in this connection, the ethical quality of its specific transactions with competitors is not considered, but only its relations with Natural Law and the general public. Its strong points as compared with other trusts may be noted: (1) It deals in a product which in some measure is a natural monopoly. Petroleum, of any amount, is the crude product of but few countries and localities. (2) It has stimulated demand by furnishing to the world good goods at normal prices. (3) Great executive ability, extensive operations, and improved and patented processes in production, all combine to give it cohesion and solidity. It has made itself a great unit, and, unlike most other trusts, is not composed of a number of smaller units held together by an artificial tie. But even this great modern phenomenon of the business world will not forever escape disintegrating forces. It will be a marvel if after the present remarkable executive management shall have passed away, its place can be completely filled.

There are some other capitalistic combinations which from peculiar circumstances, in various degrees, are natural monopolies. Typical among them may be named anthracite coal combinations, gas trusts, and the Western Union Telegraph Company. It is at once apparent, that so far as any commodity or service is exempt from *general* competition, it is not fully amenable to Natural Law; and yet, even in such cases, the public has much greater protection than is popularly supposed. Strange as it may at first seem, the actual safety-valve is the motive of gain. Take the Western Union Telegraph Company. Though not a natural monopoly, circumstances have made it a close and complete one. How can it realize the largest profit? Some would say by charging higher rates for service, but it is probable that such a policy would result in positive loss. Demand would at once fall off. If present rates were doubled, many who now make every-day use of its facilities would do so but rarely, and only for imperative reasons. It is quite probable that a reduction from present rates would enhance profits. There is a fair and normal value at which supply and demand meet, to the mutual advantage of producer and consumer. Even great corporations often mistake their true interest, and only learn by slow educational experience that their own advantage coincides with that of the public. If the price of telegraphic service were lowered one-half, only the same plant and but little more help would be required. The business might increase fivefold, and telegrams become almost as common as communications by mail. Doubtless the company would serve itself in serving the public. The working of the same law is seen in the immense increase of railroad freights, made possible by rates so low that they would have been pronounced ruinous by experts ten years ago.

Those combinations popularly known as corners deserve attention because their power for harm is so generally overrated. But while the public has little to fear from them, they are both demoralizing and disastrous to the great majority of their promoters. Where one party must lose in order that another may gain, the transaction is abnormal. Ideal commerce presupposes that both parties are gainers. No possible corner in any product can more than locally and temporarily affect the *bona fide* consumer. In a successful corner — which is exceptional — the "short seller" is the sufferer. The ambition to "control" some product, and the insane desire to acquire unearned wealth quickly, lead many unscrupulous business men to forget the great equalizing power of rapid transportation, instantaneous communication, and the general forecast of future conditions, all of which, more than at any previous time, render attempts to corner any product extra hazardous. Speculation in "futures" is the great bane of the modern business world; and yet it is impracticable to legislate against it, because genuine and speculative transactions shade into each other by indistinguishable degrees. Such legislation would also impair the freedom of individual contract, which should be sacredly preserved. Here, as elsewhere, a line must be drawn between a normal transaction and its possible abuses. The simple fact that a transaction is in a "future" is, in itself, no evidence that it is artificial or even speculative.

As illustrative of the principles of a corner, let us outline a conventional one in Chicago. Pork is selling in April for twelve dollars per barrel. There is a fair stock on hand, and many believe that in the meantime prices for July delivery will decline somewhat. They are therefore willing to sell the July "future" for twelve dollars, or even a little less, though they may not have a barrel on hand at the time of sale. They expect to buy for less before July, and thereby make a profit at the time of delivery. These are "short sellers;" and they furnish the opportunity for a corner, and have an equal responsibility for it. Another party believe that natural conditions favor somewhat higher prices, and that by buying freely they can still more enhance them. They form a syndicate to "run a corner." They very quietly proceed, through their brokers, to buy up, not only the real pork on hand, but also that "sold short," or pork to be. Both parties furnish guaranty funds called "margins" for the faithful fulfilment of contracts. The syndicate gradually advance the price by their purchases. Though they may lose something in the end on the *actual* pork, they expect to make a great deal more out of the short sellers. They crowd up the price by fractions to fifteen, sixteen, and finally to seventeen dollars before the end of July. Heavier "margins" are constantly required. The contest is wholly outside of natural conditions. The combination proceed confidently, and believe their adversaries are at their mercy. During May, June, and July, pork, naturally worth twelve dollars, brings much more, and it "pours in" unexpectedly from all directions. The syndicate must take care of it, for to retreat would involve great loss. Stocks in neighboring cities must be reckoned with, for they all gravitate towards an inflated market. It is *all* piled up; for so long as artificial prices are current, dealers and consumers in the East and Europe refrain from buying. The only market for the syndicate consists in the probable demand from the short sellers to fill their outstanding contracts. The operation becomes gigantic, and all available funds are exhausted in margins, and more called for. Collapse follows, and pork falls five dollars, or thereabouts, in a single day. The combination began with great means and expectations, but miscalculated silent forces.

The illustrative supposition just outlined was substantially paralleled by the veritable failure of a *real* pork corner in Chicago in the summer of 1893. The principles involved in all corners being practically identical, one typical case will suffice for a class.

Something more than mere magnitude must be alleged against combinations of capital to condemn them. The difficulty is not with the principle of combination, but in the abuses and excrescences that come through human avarice. The great enterprises which form an important part of our complex civilization cannot be carried forward without the herculean forces of combined capital. They are an embodiment on a grand scale of the law of co-operation. But any combination, whether or not it be called a trust, cannot violate Natural Law with impunity. If the transgression be of great magnitude the inevitable punishment will be in proportion. Retribution is inherent. The economic, no less than the physical law of gravitation is never suspended. If any combination lacks organic unity its days are numbered. Any trust forcing artificial values soon galvanizes into life new and menacing competition on all sides.

Trust combinations have been nearly or quite as common in England, Germany, and France as in the United States. So far as they embody the abuses of combination they are the outcome of cupidity and ignorance. The trust has been a popular and demagogic "bogy," but it need not be feared except by the limited number who dabble in its stock. Its bitter penalties are stored up within its own boundaries. So soon as moral and economic education becomes more general, it will be conceded that value is conferred only by inherent quality, and that combinations — whether of capital or labor — are utterly incapable of its creation. Impersonal conditions, and not combined dictation or coercion, form the basis of *how much* a thing is wanted. Attraction, repulsion, and cohesion are as regnant in the world of economics as in that of matter.

COMBINATIONS OF LABOR.

"*Every rise of wages which one body secures by mere exclusive combination represents a certain amount, sometimes a large amount, of injury to the other bodies of workmen.*"

W. STANLEY JEVONS.

"*The highest form of co-operation is all-inclusive.*"

"*The truth shall make you free.*"

JOHN viii. 32.

"*The time, however, is past when the friends of human improvement can look with complacency on the attempts of small sections of the community, whether belonging to the laboring or any other class, to organize a separate class interest in antagonism to the general body of laborers.*"

JOHN STUART MILL.

"*There is a constant danger lest the Spirit of Association should attempt to act against Nature instead of acting with it. There is, for example, a Law — an observed order of facts — in respect to Man, which the working classes too often forget, but which can neither be violated nor neglected with impunity. That Law is the Law of inequality — the various degrees in which the gifts both of Body and of Mind are shared among men. This is one of the most fundamental facts of human nature. Nor is it difficult to see how it should be also one of the most beneficent.*"

THE DUKE OF ARGYLL.

VII.

COMBINATIONS OF LABOR.

The combination of labor is proper and legitimate. There is a natural *esprit de corps* and brotherly interest among those whose occupations and experiences are alike. They feel the impulse of the same wholesome ambition, and have similar obstacles with which to contend. Man is a social being. Members of the same profession or handicraft are naturally drawn together, and societies and leagues may be formed for many laudable purposes. Social recreation and entertainment are indispensable among manual laborers, and congeniality in large measure belongs to those of similar habits and pursuits. Co-operation and fraternal interest in cases of misfortune or illness are always noble, but among comrades of the same calling they have a peculiar beauty and propriety. Organization is also useful as a regulative influence in outside relations, especially in negotiations with employers regarding hours, privileges, recreations, and sanitary supervision. There is great profit in reading and literary organizations, lyceums for debate, societies for the promotion of temperance and morals; scientific and trade associations having for their object the increase of technical knowledge in the various arts and professions, — all these, and others that might be named, are of great advantage to working men.

But in a study of labor organizations *as they exist*, we are reluctantly forced to conclude that the various commendable purposes above enumerated are largely lacking, and in their place are often installed various abuses of the principle of association. While the ideal labor union would be in the highest degree helpful, the actual and existing one is permeated with fallacious theories. It is necessary to note the short-comings of the combination, in order to evolve an ideal therefrom of what it may become. The very mistakes of the actual union have in them an educational influence which is a prophecy of improvement. There is an active evolutionary tendency which makes itself felt, and is even promoted by ferment, agitation, and adverse experiment.

In noticing some mistakes of the conventional labor union we do *not* oppose the *union*, but suggest laws and tendencies which, if understood, would transform it from what it *is* to what it should be. The *associative* principle is good, but its application is at fault. Natural Law, being normal, is truthful. The plain facts are what the laboring man greatly needs. His prejudices have been played upon to his own detriment. Those who stimulate his envy and antagonism are not his real friends. They pose as his champions, not by showing him his honorable and indispensable place in society, but by turning him against his own interest as well as that of the community. Their mistaken, though often honest efforts develop and increase his dependency so that he becomes the victim of a false philosophy, and inevitably realizes loss, both moral and pecuniary.

Let us, in order, note some of the fundamental misapplications of the associative principle.

First. Their spirit and temper are antagonistic to capital, or accumulated labor.

Second. Their influence is against the exercise of individual industry and excellence, and tends toward dependency.

Third. Personal freedom of action and contract is surrendered to the control of others, whose judgment is often faulty and prejudiced.

Fourth. They are tyrannical in their action toward all unorganized laborers.

Fifth. Their logical tendency and influence are in the general direction of a levelling coercive socialism.

Let us examine these points in order, as above mentioned.

First. Their spirit and temper are antagonistic to capital, or accumulated labor.

The idea of the necessary existence of this sentimental enmity has been industriously promulgated; and this, combined with a degree of jealousy in human nature toward those whom we imagine to be better off than ourselves, has given popular currency to this feeling. It has become such a habit to speak of the "interest of labor," and the "interest of capital," assuming that each is opposed to the other, that we adopt the practice without thinking of its unreasonableness. There is no natural antagonism, because both are mutual allies and necessary parts of one unit. When one suffers, both suffer; and when one is prosperous, both are. There is no more logic in a quarrel between them than there would be between the right hand and the left, or between two wheels of the same machine. Such a conflict is purely an invention. As well imagine a war between bricklaying and commerce, or industry and banking. Persons may disagree, but occupations, conditions, and truths, never; for they are all interdependent parts of one unitary system.

There are many leaders, agitators, and politicians whose interests lie directly in the line of keeping up this harmful and expensive sentiment. The machinery of labor organizations furnishes them with many opportunities to gratify ambition, love of notoriety, sense of power and authority, and to gain financial benefit and political capital. It is not claimed that all are influenced by such considerations. We are discussing principles, and not men. No doubt some are interested in this work who are conscientious, and sincerely feel that they are really aiding "the cause of labor." Here comes in a principle before noted. With one error for a basis, a whole group of erroneous relations are evolved to harmonize with it so as to form a system. Assuming that capital and labor *are* enemies, the logical result would be combination, offensive and defensive, with close ranks, thorough discipline, and perfect equipment for warfare.

If there were no accumulated capital, there would be no factories, mills, railroads, machinery, or wages. How can capital be our antagonist when its absence would throw us back into a state of barbarism? Without it, every comfort, luxury, and improvement would be wanting. Its enterprise enlarges every field of operation, increases the demand for labor, and enhances its market value.

The sentimental theorists who write on political economy fail to see that their teachings are contrary to the foundation principles of economic science, for the reason that their business education has been entirely theoretical. It would be an interesting experiment if some of these writers on "Labor Problems" would embark in *real* business. Let one of them take the management of a large manufacturing corporation, another the control of some railroad system, and a third assume the direction of affairs in a large importing or wholesale house. If consistent, they would conduct these various enterprises on the sentimental basis. In hiring help, they would not be governed by the market price of labor, but pay inefficient men the same as those of the best grade; or rather, perhaps, the price should be fixed by the local "district organization." They would pay ten hours' wages for eight hours' work, and employ none but union men, even if others were starving. The mercantile house would handle nothing but union goods, even if just as good non-union articles could be had for ten per cent less. The railroad manager would have no rolling-stock that was not made by unionists; and if his switchmen struck, he would not hire other good men who might apply for work at the market price. He would grant the terms asked by the strikers, and take them back, even if he knew that they would strike again the next day. If he wished to change rules, hours, or methods, he would first get permission from the nearest "member of the executive board," whether or not that official knew anything of the nature of the business; and his negotiations would be entirely with this official, and not with the men. It is very probable that in each of the supposed cases a year's trial of sentimental management would thoroughly satisfy the respective stockholders in regard to its merits, as compared with real business methods. Doubtless it would also satisfy employees, as the various enterprises would probably have to suspend, and they lose their places. It is one thing to assume business conditions suited to a theoretical treatise, and quite another to act under those conditions in real life. The cases supposed would be only putting in practice the everyday claims and theories of labor organizations. Suppose that the commerce of cities and nations were conducted on such a basis as an experiment. We can imagine that it would not continue long before both laborer and employer would cry out for another Adam Smith to lead them back from chaos to the solid ground of natural principles.

The prejudice of the wage-worker is inflamed until he feels that it is necessary to go into a strong combination for his own protection. Capital is personified to him as an unscrupulous, overbearing, and rich opponent, who is doing his utmost to crush and degrade him, and with such an impression his antagonism is naturally aroused. Thus the interests of all parties suffer.

Second. The influence of existing labor combinations is against the exercise of individual industry and excellence, and tends toward dependency.

It is self-evident that when one depends upon the organization of which he is a member to maintain or advance his wages, rather than upon his own individual merit, he is on the road toward dependency.

Good honest muscle, skill, and energy are the most staple of all things, and they rarely fail to meet with good demand. This is especially true of every one who is conscientious in regard to the interest of his employer. It seems to be the aim of labor organizations to make the laborer as inefficient as possible. Theirs is a levelling process, and any special energy is discountenanced. One who displays these qualities is cultivating independence; therefore he receives no encouragement. It is assumed that labor is a necessary evil, and the less of it one can get along with the better. The theory is, that with fewer hours, or a smaller amount accomplished, the more room will be left for the employment of others of the organization. It requires but a glance at these well-defined tendencies to see that they are unfavorable to the formation of any type of character that is manly or self-reliant. The sentimentality of the times that looks upon the workingman as a poor, oppressed, down-trodden being, is absurd when applied to an American laborer, and his self-respect ought to rebel against any such assumption.

The theory that wages are worth any specified sum, regardless of the market, is not sound; and every workman of any intelligence ought to be able to see this. There is no other possible conclusion in harmony with Natural Law, but that anything, even labor, is worth just what it will bring in a free and untrammelled market. It is impossible to devise any other measurement. The idea that wages must yield a fair support under all circumstances, is impractical in the business world. Charity is the highest and brightest of all virtues in its legitimate sphere; but its province is not in fixing market prices. The charitable idea would not harmonize with the dignity of labor, and every intelligent and self-respecting laborer would scorn the sentiment that he is a pauper or semi-pauper, or that he ought to receive what he had not fairly earned. No; the average workingman of America is well able to take care of himself, and not so imbecile as to require guardianship. It is for him to decide whether or not it is for his interest to accept an offer for his services, or to look for a more favorable opportunity. He is a man; and as such should do business for himself. As a social unit, he should strive to become an independent member of society. If in individual experience he make some mistakes, even these have educational value.

Third. Personal freedom of action and contract is surrendered to the control of others, whose judgment is often faulty and prejudiced.

The question comes to every intelligent workman: Can I afford to surrender my personal liberty, freedom of choice, duties to myself, family, and conscience, to any secret and irresponsible tribunal? Ought I to consent to be ordered "out" or "in," regardless of my personal wishes, in a land where individual liberty of action is the distinguishing characteristic?

With perhaps the partial exception of the society known as the "Brotherhood of Locomotive Engineers," which is a more intelligent and benevolent organization than the average, the system of strikes seems to be an important element in the working policy of labor combinations. Strikes are violent efforts to defy Natural Law, and are, therefore, harmful and expensive. Even when apparently successful, it will be found that their influence, in the long run, is disadvantageous.

The most conservative and moderate estimate of the yearly loss to the laboring men of the United States, caused by the strikes of 1886 and since, would mount high up in the millions; and the indirect results, if their influences could all be traced, would be even greater. The discharge of a single union man, or the retention of a non-unionist, has been the excuse for strikes involving thousands, with great attendant suffering and loss; and all this to vindicate a supposed principle, which really turns out to be only a sentimental "boomerang." Such was the nature of the very extensive and disastrous strike which took place in the spring of 1886, on what was known as the Gould system of railroads, extending from St. Louis to Texas. Thousands of men, many of whom had families depending upon their daily earnings, were "ordered out" of good situations, which they never afterwards regained. Thousands of others, whose occupations were more or less directly connected with these men, were thrown out of employment and business of all kinds largely interrupted for weeks in three or four States, causing great loss to all classes. Much property was destroyed, and many non-union workers injured and maltreated. All these, and other ills too numerous to mention in detail, resulted from the ill-advised and cruel orders given to confiding men by labor officials. It is a fact beyond doubt, that what would have been a great and general resumption of prosperity in the business of the whole country, which had begun to set in strongly in the early spring of 1886, was not only postponed, but almost destroyed, by the labor disturbances which came in April and May of that year, like an epidemic. The same thing has been often repeated since. And furthermore, these troubles were not spontaneous in their character, but were "ordered," engineered, and fomented by "agitators," who did not belong in the ranks of the workingmen. If things could remain in their normal and peaceful condition, so that an era of general prosperity could once get under way, the increased demand for labor would cause a *natural* advance in wages and general prosperity. One more example of "killing the goose which lays the golden egg." Natural

Law is a most powerful and serviceable friend; but if we persist in its violation, we must reap the consequences.

One or two more instances of the effect of strikes will suffice, for they are all quite uniform in their results. In July, 1886, a large number of tanners, employed in the extensive establishments of Salem and of Peabody, Mass., were "ordered out." In the end, most of the men lost their places, and had to remove elsewhere to find work. During several months, assaults, intimidation, and disorder continued, and these towns were put to large extra expense to protect, as well as possible, those who wished to work. The indirect losses and suffering growing out of this strike can never be estimated. The Peabody *Reporter* gave a very careful estimate of the direct results on Nov. 10, 1886. A personal canvass of all the shops of Salem and Peabody was made, and every detail carefully ascertained. It reported as follows:—

"On July 12, 1,500 men left work in forty-three factories, and on November 10 there were employed in these same factories 1,205 men. In the other thirty-one factories, 613 men went out at the same time, and in these factories 500 men are now at work. This makes a total of 2,113 men men who quit work July 12, and a total of 1,714 men employed in the same factories to-day. Had these men worked, they would have received $456,408. It is estimated that they have lost, aside from the amount received from the Knights of Labor, $304,272."

As a result of the eight-hour agitation during the spring and summer of 1886, the pork and beef packers of Chicago gave that system a continued and thorough trial. As competing cities continued on the ten-hour basis, the inevitable result soon became apparent. The business could be done more cheaply at those places, and, as Natural Law is never idle, the industry was rapidly transferred to them. The Chicago packers, finding it useless to hold out against the inevitable, notified their help early in November that they would be obliged to return to the ten-hour system. Rather than accede to this, twenty thousand men were "ordered out;" and this just at the beginning of winter, when a great majority had nothing ahead, and thousands had families dependent for subsistence on their daily labor. The hostile presence of such a mass of idle men made it utterly unsafe for any minority to continue at work. The few who attempted it found their lives and homes in imminent danger. Plenty of non-union men, who wanted employment, could only be scantily protected by two regiments of infantry, sent at the expense of the State, in addition to a large force of private police. It is true that officials of the Knights of Labor ostensibly discourage violence, but the difficulty is with the system. It is a cruel thing to order twenty thousand dependent and unintelligent laborers out of their positions at the beginning of winter, and it is folly to expect that they will stand idly by and see their places taken by others. It is farcical to say to them, "keep quiet," for these unfortunate men have a terrible pressure on them, forcing them *not* to keep quiet.

Look for a moment at the ultimate effect of a rise in wages, when caused by the pressure of labor organizations, without striking. For illustration: Suppose that the natural and competitive cost of a certain style of boot is five dollars per pair as produced in Lynn, and that one hundred thousand pairs are made and sold annually. Now suppose that the labor unions in that city get such a thorough control that by quiet pressure, the crimper, laster, stitcher, and all the various other kinds of workmen each establish a moderate advance, so that it then makes the cost of the boot five dollars and twenty-five cents instead of five dollars as before. The combination has carried its point, but has it made any gain? The first effect would be felt in a lessened demand. The average man would wear his old boots a little longer, or buy some other style in place of them. Instead of one hundred thousand pairs, it would be found at the end of the year that a less number, say ninety thousand pairs, had been the limit of demand. Therefore one-tenth of these workmen have been thrown out of employment. Now look at another tendency. If in Haverhill, and other competing points, natural competition still enables the same boot to be produced for five dollars, the business would at once begin to leave Lynn; for, by Natural Law, *it always seeks the cheapest producing points.* Not only the general demand would fall off, but competition would soon force the Lynn manufacturers to stop entirely the production of this boot. Some of their workmen would have to sacrifice their homes, and move elsewhere, and that city would decline in prosperity. Some one may reply that by general combination, the advance could be obtained at all points in the State. That would not in the least affect the first result, which was a lessening of general demand. But, in addition, if all sections of the State combined, it would tend to drive the business to other States, to the West and other parts of the country. This would cause a loss of dollars to the Massachusetts bootmakers, in an attempt to grasp dimes. With the levelling influence of world-wide competition, such artificial coercion is only harmful. The principle illustrated in the case of boot manufacturing is universal in its application, and no kind of production is exempt from its irresistible control. A man might as well try to lift himself by the straps, when wearing a pair of these boots, as to expect to mount above the force of these fundamental business laws, or to escape from the penalty of their violation.

If it were possible by universal combination among working-men to advance wages fifty per cent, it would not in the least improve their condition. The price of everything which they need would be enhanced in the same proportion, and they would have no larger surplus at the end of the year than they had before.

The system of assessments necessary to keep in motion all the machinery of labor organizations, including the salaries of officials, together with the very large sums necessary to maintain in idleness those who are "out," add still more to the burdens of the working man. War is always expensive; and this conflict, not with employers, but with supply and demand, is a costly operation. The promises made by the labor agitators seem attractive and desirable, but their fruits turn out to be bitter. It is noticeable that vehement champions of the labor cause, who have been zealous to have the wrongs of the working man righted through coercion, have soon after, in many cases, been found in the field as candidates for some political office.

When sentimental agitators try to make the working man restless, by teaching him that labor is a dependent condition, and that he is in "slavery," he should not forget that the average price of labor for many years has been steadily advancing from natural causes. It is now about double what it was thirty or thirty-five years ago. On an average, it took more than twice the amount of labor to buy a given amount of flour, sugar, clothing, and most other family necessities, than it does at the present time. The single item of rent is perhaps dearer; but with that exception, nearly every necessity and luxury has declined during the period in which wages have doubled. This shows a great advance in labor values from the operation of Natural Law; and they would have been still higher than they are had their rise not been obstructed during the last few years by the detrimental operation of labor unions. In spite of the influence of unrestricted immigration, the general rate of wages is more than double what it is in Europe. Everything shows that the present unrest, now existing among the manual laborers of America, is in no degree the result of changed conditions for the worse; but that it is due to a false philosophy, the seeds of which are persistently sown by foreign agitators, whose theories are advertised so abundantly by the sensational portion of the newspaper press. The vagaries of the greatest extremists thus get a large amount of notoriety.

In regard to boycotts, it is observable that they are unbusiness-like and revengeful in their conception, unnatural and un-American in their methods, and deranging in their effect on all legitimate business. It is a privilege and a necessity for the wage-worker who has limited means to expend, to buy the best goods at the lowest prices possible in a free market, whether or not they were made by members of a union. To pay more for purposes of revenge is a loss to the laborer, and an injury to society, of which he is a part.

Fourth. Labor combinations are tyrannical in their action towards all unorganized laborers.

They assume to represent labor in general, but statistics show that but a small part of the grand aggregate of laboring men belong to organizations. The interests of this large majority are, to a great extent, ignored by the public, and despised by the combinations. No matter how honest, industrious, and law-abiding they may be, they are "scabs," and receive moral, and often physical treatment more worthy of criminals than law-abiding citizens. The public, with a singular absence of that sense of justice which is theoretically dear to American citizens, seems to expect more or less of this condition of things as a matter of course, and moral and social abuse, when not accompanied by physical assault, is treated as a matter of slight consequence. Legislative law professes to protect every man in his right either to buy or sell labor or any other thing of value in the open market. A recent writer has well said that "attempts to do away with this right by force, intimidation, or interference, have their logical end in *anarchy*. The majesty of the law is the foundation of all liberty and prosperity, and every man should give it his moral support."

Sentimental writers, as a rule, utterly ignore the great unorganized majority of laborers, as if no such people existed. When they speak of labor, they refer only to the minority portion, or that which is organized. Are not these men American citizens? and are they not entitled to common rights and protection under a form of government which professes to be democratic and impartial? They have a right to sell their services as they choose to willing purchasers, and when the government fails to protect them in this privilege, then the boasted American freedom is a farce. These men, as a class, are ignored by the politician in his zeal to bid for the labor vote, and even the newspaper press, as a rule, gives them scanty recognition. They are peaceable, law abiding, and unobtrusive, but at the same time form a very important part of the live-oak in the hull of the "Ship of State."

Fifth. The logical tendency and influence of labor combinations are in the direction of a levelling coercive socialism.

Socialism as a political system will be elsewhere discussed, but the logical tendencies which have cropped out of the agitations of organized labor are so marked that they may be briefly noted in this connection. When natural and business principles are left behind, and sentimental methods adopted, all solid ground is abandoned. As well attempt to found a solid structure on the quicksand, or combine mathematics with fiction, as to expect that business can prosper with personal independence and ambition crushed out. The natural and logical outcome of any compulsory kind of socialism, in the end, is the disruption of society and government. There is a close and growing sympathy between labor and socialistic organizations, especially in the larger cities. The socialist holds out an alluring bait to the ignorant masses of foreign laborers, and soon they are made to feel that because some others have more of the results of accumulated labor than they possess, the world has not been fair with them, and they have not had their "rights."

The only test of the soundness of *theories* is contained in their practical working, and this renders the experience of Australia with organized labor of interest. Nowhere else on the face of the globe has unionism ever gained such a complete domination as it there possessed previous to its recent dethronement. In the *Engineering Magazine* for April, 1893, Edmund Mitchell, an able writer and economist of that country, gave a detailed account of the great contest. It finally disrupted the unions and also nearly wiped out the principal colonial industries, turning prosperity into chaos, and this notwithstanding their almost unbounded natural agricultural and mineral resources. Four long and desperate strikes extended into trades, occupations, and localities, entirely distinct from the original controversies, involving thousands of innocent people in distress and bankruptcy. Among many other interesting particulars of this long industrial war, Mr. Mitchell says: —

"It has to be noted that in no single instance did these disputes originate from or hinge upon a disagreement as to wages. Brushing aside a few minor issues involved, we find that the one cause of quarrel throughout was the demand on the part of the strikers for the exclusive recognition of unionism, and the firm determination of the employers to refuse to concede that demand. Had the unionists won the day, there can be no doubt that every worker in Australia earning his living by the sweat of his brow would have been compelled to join one or other of the labor organizations, and place himself under the domination of the small cliques of individuals in the big cities who make of labor agitation an exhilarating and lucrative profession. To show how thoroughly labor militant in Australia has forced employers to combine in self-defence, let us take the great wool-growing industry, which adds yearly to the wealth of these colonies from £20,000,000 to £25,000,000. The lethargy and lack of cohesion among pastoralists enabled the shearers' union, three years ago, to acquire a position of almost despotic power. Its leaders boasted that they controlled the shearing in ninety-five per cent of the wool-sheds. In the framing of the rules which regulated in every detail the manner in which the shearing was to be conducted, the employer had no voice whatever; he had either to accept them or to enter upon the almost impossible task of fighting the whole union single-handed. Every shearer was compelled to take out his union ticket, paying, besides entrance fees, £1 per annum for the privilege; and the man who refused to submit to this blackmailing process was declared a pariah, by whose side no unionist would work or eat, was hounded from wool-shed to woolshed without the chance of securing employment, and was finally driven out of the industry. The weapon of the boycott was ruthlessly used against employers and non-unionist shearers alike; and some conception of the wide-reaching development of the system will be derived from an extract from a manifesto issued by the shearers' union: 'We intend,' ran this official document, 'to teach the squatter the folly of resistance to our combination. He shall not be allowed to shear his wool except by union labor. But if he should succeed in getting the wool off the sheep's back, it may rot in his sheds, for we shall prevent the carriers taking it to the railway; and should he succeed in getting it to the railway, we shall prevent it going to sea, for we shall call out the sailors and the officers; and if it sails, we shall prevent its discharge in London, for we shall call out the dock laborers.'"

In speaking of the Broken Hill (Queensland) mining strike, Mr. Mitchell observes : —

"The leaders, who are now serving sentences in jail, showed themselves to be professional agitators pure and simple. Possessed of the gift of fluent speech, these men, not miners by calling at all, had foisted themselves upon the workers' associations, and by the rhetorical trick of inflaming envious passions and stirring up strife between the employers and the employed, had soon attained to positions of personal ascendency, the toleration of which among large bodies of fairly-educated, self-respecting workingmen is almost incredible. The strike was the very opportunity desired by the leaders. At one bound they became persons of public importance, issuing fierce manifestoes, having their speeches telegraphed across a great continent, visiting their pickets like generals in the field, being huzzaed by the mob as they passed along the streets, and generally living in a constant vapor-bath of self-esteem and servile flattery. All these are simply the necessary preliminaries to securing a seat in Parliament, and what to a man of the working classes is a very large income, £300 per annum, with no real hard work to do, with free railway travelling and invitations to official dinners."

The fundamental issue in the Australian, as in all other attempts at unionist dictation, is: *Shall the freedom of contract be destroyed?* Out of a total population of several millions in Australia there were only seventy-five thousand unionists. Can there be any justice in the claim of so small a fraction of the able-bodied male population to monopolize the whole labor market? Can they in any sense be said to represent the *interest* of LABOR?

The abnormal state of affairs in Australia brought on a financial panic, and upset all industrial enterprises. Investors will not invest but withdraw their means, so far as they can, wherever there is chronic friction and unrest. Australian progress was turned back, and years will be required for it to regain its former momentum and again induce such a return of confidence as will attract capital and develop resources. Privates in the ranks of labor are often unconscious of the tyranny that is perpetrated in their name because it is claimed to be for their interest.

The interests of employer and employee are one, and it is to the advantage of both that there should be mutual confidence and sympathy. The more conscientious and hearty the service the more the employer can afford to pay for it; and the more, on an average, he *will* pay. Half-hearted service will not permanently command a high price. The union official strives to rend asunder the two elements which form the natural unit, and, in the degree that alienation takes place, both are injured—the employee the most.

Any schism among members of the social organism is a common misfortune. Confidence in the stability of business conditions is the life-blood of prosperity.

All inefficient or indifferent service causes moral decay in the character of the laborer. Even in abnormal cases where his remuneration is less than the current rate he cannot afford to degrade his manhood. The theory that in the aggregate there is a certain fixed amount of work to be done, and that fewer hours and less accomplishment will give more workmen an opportunity, is shallow fallacy. With peaceful conditions and prevailing confidence the ever-expanding demand of the world equals any possible supply, and this even with all the labor-saving machinery that has been or will be employed. It may be asked: Do you favor long hours? No; but personal freedom. If one choose to work ten hours instead of eight it is his privilege, and no man, organization, nor even the State, has the moral right to coerce him. The hours of labor are growing less from natural causes. The workman *does* need time for mental and moral improvement, but, important as these are, freedom is still more so. The sagacious employer, even from the sole standpoint of pecuniary success, will find it to his interest to shorten hours, and favor his help in every way that surrounding conditions will possibly allow. Natural evolutionary progress is in the direction of shorter hours, but there should be no arbitrary dictation. The wise employer will respect the manhood of his employee and keep up his *esprit de corps*.

The seeming over-supply of labor comes from obstructive dictation and impaired confidence. The capitalist who would build a block of houses will hesitate long before beginning, if he is likely to be harassed by strikes, boycotts, and the walking delegate. Business and confidence can no more grow under such conditions than could a garden flourish were it continually trampled over in a disorderly manner.

In some departments an overstocked labor market is the result of a baseless fancy as to the relative social grades of various kinds of service. For illustration: Among the occupations of women, the market for good domestics, in excellent positions, at good wages, is never overstocked. But shirt makers, who might make good domestics, will work at starvation prices in attics, because the latter are regarded as socially inferior. But all genuine inferiority is located only in character. The sentimental economist seems blind to the fact that false pride, ignorance, and improvidence — to say nothing of intemperance and crime — are responsible for the over-supply of labor, and goes out of his way to lay the blame upon "competition" or a wrong "social system." He makes a complicated "problem" out of that which should be plain to the commonest commonsense.

The law of compensation is untiring in finding the specific gravity of every person, and in meting to him his deserts. If it seem to fail in some cases from the standpoint of mere monetary accumulation, it will not permanently vary when tested by the truer measurement of human harmony and happiness. These are popularly supposed to be secured only in financial profit, and therefore wealth is earnestly sought. A deeper view, however, proves that mere pecuniary success is but the lower and smaller part of them.

If pebbles could be coined into money to support every inefficient and improvident person, it would do him an injury, for he would miss all educational influences and disciplinary penalties. It is in accord with the purest altruism to declare that the artificial removal of the corresponding punishments which are the natural fruit of the conditions before named would be positively uncharitable. If the great evolutionary force which pushes men toward higher character were wanting, progress would be paralyzed. The scientific way to relieve suffering is to reach it through its causation. The ills of society are the harvest of defective character, and it is as logical to refer them to lunar influences as to the social system, land system, or the law of competition. If ignorance and laziness were not logically followed by want, they would never be outgrown. Beneath all the seeming severity — which does and should enlist our hearty sympathy — natural penalties are kindly. But intelligent philanthropy will address itself to the underlying causes. Sentimentalism cannot improve upon, or reverse, the divine plan of evolutionary progress. It is prolonged cruelty to assure men that their trouble is external. The charity to bestow is industrial education, self-help, faithfulness, honor, temperance, love, *character*. To misplace the fault is to be unkind to the individual and to society. If one leans upon anything outside of his own talents and powers, he is leaning down hill.

The restrictions put upon apprenticeship and industrial education by conventional unionism are distinctly reactionary and harmful to society in general. The most helpful, hopeful, and important agency for the cure of prevailing economic ills is general *industrial education,* and this not alone for the male sex. The education of girls in household economy, which is becoming more general, is a great advance in the right direction.

The *ideal* labor union would strive to make its members experts in their respective vocations. It would inspire them with wholesome ambition, independence, and honor. It would educate them technically and morally. It would make them MEN. It would fit each one to rise through natural ambition and merit to the rank of employer. "*There is always room at the top.*"

Men are not working for some intangible despot called Capital. Both capital and labor are impersonal conditions; while all injustice is personal.

COMBINATIONS OF LABOR. 95

As a rule the employer hires his capital and the capitalist is not an employer. If either violates Natural Law the penalty is inevitable. Is this doctrine in accord with the golden rule or law of love? Assuredly yes. If sin, economic as well as moral, did not bring penalty in its train, men would sin forever.

The principles enunciated in this chapter are in the interest of laborers, and from their true standpoint. In proportion as truth is recognized prosperity will be the rule, wages advance, and confidence prevail.

Prejudice and antagonism invariably bring forth bitter fruit, and this rule finds no exception, in any rank or condition.

EMPLOYERS AND PROFIT SHARING.

"But he whose inborn worth his acts commend,
Of gentle soul, to human race a friend."
                                        POPE.

"And each shall care for other,
And each to each shall bend,
To the poor a noble brother,
To the good an equal friend."
                                        EMERSON.

"Why should a man whose blood is warm within,
Sit like his grandsire cut in alabaster?"
                                        MERCHANT OF VENICE.

"And learn the luxury of doing good."
                                        GOLDSMITH.

"O slavish man! will you not bear with your own brother, who has God for his father, as being a son from the same stock, and of the same high descent? But if you chance to be placed in some superior station, will you presently set yourself up for a tyrant?"
                                        EPICTETUS.

"You cannot do wrong without suffering wrong. Treat men as pawns and ninepins, and you shall suffer as well as they. If you leave out their heart, you shall lose your own."
                                        EMERSON.

VIII.

EMPLOYERS AND PROFIT SHARING.

NATURAL LAW is as supreme and unrepealable, and its penalties as sure, in the realm of capital as of labor; therefore its dominant tendencies hold as firmly with employer as with employee. The sentimental alienation now existing between "capital and labor" is by no means entirely due to the unreasonableness of the latter, the spirit of unionism, or the machinations of agitators. The unsympathetic hardness of the employer furnishes the soil where coercive unionism takes root and thrives. Abuses, like trees, never grow entirely one-sided. An abnormal protuberance in any direction indicates that there is some opposing correspondence. The fault of a creaking wheel is often found in its bearings.

In production the natural unit is the combined employer and employee. Completeness comes from the joining of unlike elements, for each supplements the other. All unions of employers with employers, and employees with employees, except for social and educational purposes, are unnatural. Superior and economical production is secured by a harmonious blending of the different parts which make up a whole. A horse and cart, being a unit for their appropriate service, can accomplish more than a hundred horses and carts when separated, or even disagreeing.

Before the inauguration of the factory system, when production was carried on under more primitive conditions, employer and employee were comparatively near together. But modern extension and specialization have multiplied the number of wage-earners under each single control, and put them at a distance from the proprietor. Corporate production also renders the welding of the parted unit still more difficult. Unless the chasm can be bridged, society loses and both parts suffer. How can the employer most effectually harmonize prevailing discord? If it be impossible for him to maintain the personal nearness to his help which was formerly the rule, something else equally good must be substituted.

While business should be done on business principles, there is abundant room and opportunity for other obligations outside of that of mere service rendered and paid for. Natural Law comprehends within its scope not only economic rules and methods, but it also provides an important place for the exercise of the kindly elements that are inherent in man's nature. These, while not strictly entering into a business contract, surround and refine it, lighten its burdens, and soften its cares. They are like the springs and cushions to a carriage, which, while they have no direct relation to speed or distance, render progress much more comfortable and easy. Natural Law is democratic. It recognizes a man as a *man* so long as he fulfils the conditions of manhood. The duties of an employer to his workmen are discharged with the payment of stipulated wages, so far as they relate to the business and economic sphere; but there are other relations that cannot be ignored. They involve a recognition of the fact of man's intrinsic brotherhood, and that each individual is a part of one moral and social economy; and these relations, though on another plane, are equally natural and necessary. As harmony with Natural Law always lends powerful aid in the achievement of success, the employer who heeds these higher claims, more fully discharges his obligation to society, and at the same time smooths the road toward his own prosperity.

Employers should not forget that laborers are *men*, not machines. A larger consideration toward, and interest in, employees would largely dispel the illusion of a natural antagonism, on which labor unions flourish and production decreases. The workmen are the staff of the employer. A general might almost as well expect a successful campaign with his staff selected from the hostile army, as an employer expect good, honest service from men whose feelings are antagonistic, whether with or without good cause. Cultivate friendliness and sympathy with employees, not by flattery, but by genuine interest in their welfare. There is too little personal contact and community of feeling. Misunderstandings and difficulties vanish when discussed face to face in a conciliatory spirit. Show your workmen that you are more truly their friend than the labor agitator who comes from the outside to stir up strife, and the latter's occupation will be gone. In this direction, and this only, can the remedy for labor troubles be found. Disband the horizontal and combative combinations of laborers with laborers, and employers with employers, and cultivate alliances and interests in the other direction. This can only begin with some conciliation on both sides; for both have been looking in the wrong direction, and emphasizing a selfish independence. The pecuniary success of both parties can only be increased by some such means. This change of front is very important, notwithstanding it is contrary to the position taken by many recent writers on the "labor problem," the burden of whose effort has been to urge working-men into combinations detrimental to their own interests. All such teaching renders the solution of the "labor problem" more difficult. The head and hands must have one object, or else there will be friction for both. All that has been recommended can be done by the employer without injuring his own self-respect or that of his employee. A unity of interest between employer and employed is natural, because there is no competition between them. *Competition is always horizontal, or on the* same plane. The natural competition of employees is with employees, and of employers with employers. The union should be in a direction to form the unit. To solidify and strengthen one element to the neglect of the other, is like sharpening one blade of a pair of shears when its companion is useless.

The best employers naturally attract the best help, and such a combination has great strength. The employer must assume the risks and contingencies of his business; and if he be wise he will cultivate all those elements which tend to harmony and, as a natural consequence, to success. Suppose that after inventories are taken at the end of the year, the owner distribute a certain part of his surplus to his faithful help; even from a business standpoint, would it not be a good investment? While not a legal obligation, it would not be a charity, but simply a proper reward for special faithfulness. Can we doubt that such a course would be mutually beneficial in the long run? It would require very strong inducements to organize a strike among workmen dealt with in such a spirit. An early and notable example of voluntary profit-sharing was given by Leclair, a French employer, more than fifty years ago. After suffering from the effects of discontent, antagonism, and unfriendly suspicion among his men, he resolved to try an experiment. In 1842, after calling together the most faithful of his help, forty-four in number, he threw upon the table a bag of gold containing twenty-three hundred and seventy-five dollars, distributing to each his share, averaging over fifty dollars per man. This was an object lesson that had a telling effect. Distrust was replaced by confidence, and a friendly interest and trust became the rule. When the men found that they had an interest in their employer's prosperity, they became faithful to every requirement, and performed each duty more carefully and thoroughly. The mutual benefits of the principle were so apparent that M. Leclair formally adopted it; and although he died in 1872, his successors still continue the practice. At the present time there are more than two hundred firms in Europe which have adopted this plan substantially, varying it only in minor details. Quite a large number of American companies and individuals, also, have employed methods which are similar in spirit and practice. Messrs. Lorillard & Co., of New York, recently distributed in one year sixteen thousand five hundred dollars among their help as a part of the profit which they were willing to relinquish to their faithful workmen.

There seems to be nothing else at the command of the employer so promising, aside from social and moral influences, as profit sharing in one or another of its phases. Adequate pecuniary recognition must be given to special fidelity and length of service. The particular method of the applied principle will necessarily depend upon the character of the business. The employee must become convinced that the success of his employer includes his own success. Under profit-sharing, as already outlined, it is obvious that in cases of exceptional loss the burden must be borne entirely by the proprietor. The plan suggested leaves the full legal and moral control of everything, over and above the stipulated wages, with the employer. But special liberality will promote his prosperity, and accord with Natural Law, even though he may be unaware of it. Profit sharing is not charity, even if purely voluntary. The independence and self-respect of the employee must be preserved. The best principle is the best policy. For railroad and other large corporations the enhancement of the workman's interest may perhaps be stimulated by insurance during employment against casualty or illness, or by a system of added percentages to ordinary wages, increasing somewhat for every continuous year of faithful service.

There are still other modifications of the principle, some of which are more distinctively co-operative. One of these has been very successfully tried for several years by the N. O. Nelson Manufacturing Company, of St. Louis, of which a brief account may be of interest. Their experience has been very significant. For eight years after paying annually six per cent dividends to the invested capital, they divided an average of eight per cent dividends on wages also. Employees are also allowed to become shareholders in the company. During the financial depression of the summer of 1893, the employees willingly worked full time on three-quarters pay, for the double purpose of husbanding resources and joining in the probable loss of that exceptional year. The amount deducted from wages was to be made up out of future profits, and the capital shares, in any finally ascertained loss, in the same ratio as the wages. This is genuine co-operation, but theorists will please note that the legitimate competitive element is still present in the relations of the company with neighboring companies in the same line of production. Under such a plan employees become real partners, and their interest in the success of the company is greatly deepened.

Another plan, sometimes practicable, is to impartially merge a business into a stock company and allow employees to acquire shares at a normal valuation.

We advise all large employers, whether or not subject to "labor troubles," to thoroughly test some one of these devices for a consolidation of interests. They are in full accord with both social and economic law and promise well. PROFIT SHARING *embodies the spirit that will furnish the key to labor problems.*

Its denunciation by niggardly and short-sighted employers, on one hand, or selfish professional agitators on the other, cannot shake it, for it is founded on justice and humanity. Unselfishness should be the motive of the employer, but even from the lower stand-point to *share* profits is to *increase* them.

The employer should take a deep interest in the dissemination of correct principles in morals, temperance, and hygiene among his workmen, and by his influence and aid further all practical movements for their improvement. Opportunities for this vary much in different places and conditions, but there is room for a great and general advance in these particulars. Large employers whose establishments are in small factory towns, have it especially in their power to accomplish much for the good of their help, without any sacrifice of independence on the part of the workmen. Perhaps the most notable experiment of this kind that has been tried in the United States is in the town of Pullman, near Chicago. As the Pullman Company owned the land from the start, they were able to exercise more perfect control than would often be possible; but still, their plan might be approximated in many cases, and with great benefit. Though several thousand men are employed, no places for the sale of liquors are allowed. This alone secures, in general, a superior class of workmen. The houses for the occupation of the employees are built with careful regard for sanitary excellence, and in addition, are models in their tasteful and modest architectural effect. The water, gas, and sewer systems are of the most approved kind, and owned by the company. A public library, schools, churches, and a suitable place of amusement, receive such aid and oversight from the company as will insure their maintenance and efficiency. The rentals of the workmen's homes are moderate, being only sufficient to pay a fair interest on their cost, and other facilities are furnished for economy and comfort in living. While the workmen pay for everything they have, thus preserving their independence, they are able to get the best at low rates. The Pullman experiment has been very successful, and is worthy of imitation.

Many employers mistake their own interests, and add to their difficulties, by an unnecessary severity toward their employees, and the exercise of an overbearing and tyrannical temper. Such a spirit is a formidable obstacle to success.

What are known as "lock-outs" are sometimes resorted to, to force concessions from employees. They are unnatural, and in many cases cruel in their effects; and, except in rare instances to counteract wholesale dictation, they are reprehensible. When used for the purpose of artificially putting down the price of labor, they are to be condemned from a moral point of view; and they also bring their own legitimate punishment, as a violation of Natural Law. Any kind of combination among employers, having in view a compulsory reduction of wages, or harder conditions, is unwise, because it arouses an antagonistic spirit among employees, and is unprofitable also in its after effects. Only in exceptional cases, to resist wholesale tyranny on the part of labor unions, on the principle of combating one evil with another, can there be any excuse for combinations among employers.

What is called black-listing is also a weapon that should be used with extreme care, if at all, because it is very liable to abuse. If it were always confined to bad employees, so proved, it might have redeeming, and perhaps wholesome features. It is, however, so often employed to gratify personal prejudice, that its legitimate use is extremely restricted.

The natural elements tending powerfully towards success to an employer of labor are the development of an *esprit de corps* among his help, and the secure possession of their respect and good-will. An ideal establishment is one where employer and employee are each proud of their connection with the other. Such a combination means the highest wages, and at the same time the best and most economical production.

EMPLOYEES: THEIR OBLIGATIONS AND PRIVILEGES.

"No man is born into the world whose work
Is not born with him.   There is always work,
And tools to work withal, for those who will;
And blessed are the horny hands of toil."

JAMES RUSSELL LOWELL.

"The brave man carves out his fortune, and every man is the son of his own works."

CERVANTES.

"The people never give up their liberties but under some delusion."

BURKE.

"Be satisfied with success in even the smallest matter, and think that even such a result is no trifle."

MARCUS AURELIUS.

"Never esteem anything as of advantage to thee that shall make thee break thy word or lose thy self-respect."

IBID.

"Men my brothers, men the workers, ever reaping something new."

TENNYSON.

"The sleep of a laboring man is sweet."

ECCLESIASTES.

IX.

EMPLOYEES: THEIR OBLIGATIONS AND PRIVILEGES.

As the duties and interests of employees are touched upon in other connections, a direct study of them may be concise.

The prevailing spirit of the times tends to make the employee regard his employer with some degree of jealousy, if not of antagonism. Such a feeling is both morally and economically unprofitable. The employer is his natural supplement in production, and besides, no one else is in a position to do so much for him. Deprived of the co-operation of his employer's talent, capital, and executive ability, he is weak and incomplete. No man can afford to pick a quarrel with his productive partner, nor to antagonize the very things which are in the line of his laudable ambition. Without employers there could be no employees. The theory that production is solely the result of physical labor, as urged by some socialistic agitators, is unmitigated fallacy. If material force be all that is required, then steam, electricity, horses and mules, and even water-power, should receive the main credit, leaving the human element quite unimportant. Matter can never bear comparison with mind, nor a mountain with a man. A high order of executive talent is more rare than a corresponding quality of muscle, and therefore it always brings a higher price. Its superior value is not due to fashion or fancy, but to demand. The worth of muscle also depends upon the quality of mind mixed with it. To lay brick requires a larger percentage of mind than to dig ditches, and it therefore readily brings a higher price. By the socialistic theory that the value of products should be abstractly estimated by the number of "labor-hours" put into them, the expert is no better than the ignoramus, and the latter might as well remain what he is. Under free conditions Natural Law never makes a mistake in weighing values.

Education, moral, economic, and technical, is the great need of the wage-worker. These furnish the only solid basis for wages and for their increase. Obstruction and friction invariably tend toward their diminution. Real education is not the acquisition of a certain amount of "book learning," but the art of fitting well for present and prospective duties. It is entering upon a road which always leads higher.

Evolution is a universal law; but if one waits for it to push him from behind he advances slowly, and with friction. Attracted from before, he makes rapid progress. Excellence, as an ideal, furnishes an ever-present stimulus. If one does his best, it does not so much matter where in the great field of human effort he may be to-day, for he will soon leave the locality behind him. Some one may suggest that opportunities are important, but they are largely made, not found. "Luck" comes to those who win it. Chance appears upon the surface, but a deeper view shows that all success comes by *law*. A trade, profession, or character is like an edifice; every brick must be put in its place, and it will not get there by luck.

Work is not a curse or a thing to be avoided. Rightly regarded, it is education and development. To count it as drudgery is to gratuitously make and extract a slavish influence from it. Looked upon as an education and stepping-stone, it is refined and ennobled, and this even if low conventionally in grade. To idealize one's present vocation is to prepare for a higher and more profitable one.

Work is not to be dodged, but transformed into development. The man must lift his effort, and not allow it to become merely mechanical. One's attitude toward it determines what it is — to him. It also indicates whether or not he will advance to the rank of employer.

The enjoyment of a vacation often causes a workman to feel that its indefinite continuance, were it practicable, would be desirable. But only because work is the rule does recreation possess sweetness. To a living, progressive *man* enforced idleness is a torture. Lack of occupation causes decay. Even when underpaid, the wage-earner, out of regard for himself, cannot afford to do less than his best. Through the law of compensation every one, at length, gets exactly his due. Specific gravity applies not solely to fluids, but is universal.

The employee naturally and rightly wants increased wages; and, through Natural Law, the only road to them is to earn more or better. If he get them through the seeming short cut of coercion or organized pressure, they will soon slip back. Progress, to be solid, must be natural.

It is often supposed that employers might pay whatever wages they please, regardless of the market; but competitive relations in innumerable directions do not permit it, and, as elsewhere shown, general competitive laws are as indispensable to wage-workers as to society at large. If labor-unions, instead of limiting apprenticeship and encouraging idleness, — under the delusive theory that the total amount of work to be done is limited and fixed, — would educate their members, there would soon be enough for all to do. Each kind of labor is the patron of all the others, so that all may be increased by balanced growth. Obstruction beginning in one trade is reflected and re-reflected in all the others until all suffer. With complete harmony, wages would soon rise from enlarged demand. This is the only possible basis for any increase. With unobstructed prosperity, invention and art would open new avenues of employment, and over-production be unknown. When the sewing-machine was invented no less seamstresses were needed, because the demand for sewing at once increased to the full capacity of the improvement. In every department, with the increase of facilities, embellishment and complex design take the place of plain crudeness. Fifty years ago only the carpenter, mason, and painter were required to construct an average dwelling, while now a score of different trades lavish their skill in perfecting its convenience and adornment.

General economic confidence can only exist upon an adequate foundation. It especially requires a sound currency, an unwavering tariff, whether high or low, and, most of all, harmonious labor conditions. With this combination, employment, for every one who desired it, would be sure and remunerative. Things to be done would multiply faster than hands to do them.

The goal of the workman is to become a proprietor, and under all normal conditions such an ambition is laudable. In our own country no individual is crystallized into any fixed class or grade; and if artificial dependence, coercion, and the levelling influences of unionism, could be put aside, workmen would have an unobstructed road to progress open before them. If, however, in some cases it be impracticable to rise in grade, advancement in quality is possible, which is almost equally important.

With but rare exceptions, the most eminent and successful business men of America started in active life as wage-workers, and their secret is that in every position they did their best. This was their self-education, and education is capital. Even when under-paid that kind of capital is always augmenting. Where wealth has been acquired by dishonorable means it will prove a curse. The moral penalty for Natural Law violated is inherent, and is inevitably realized sooner or later.

EMPLOYEES: THEIR OBLIGATIONS AND PRIVILEGES. 113

However much specious theories may prevail, as to short hours, lessened production, limited apprenticeship, and much leisure, individual merit will always remain the sole basis of value for service. Economy, energy, and excellence may be popularly regarded as antique, but the link that binds success to them can never be severed. Cause and effect are connected by a divine welding.

GOVERNMENTAL ARBITRATION.

> "*What mighty contests rise from trivial things!*"
> — POPE.

> "*And sheathed their swords for lack of argument.*"
> — SHAKESPEARE.

> "*Beware
> Of entrance to a quarrel; but being in,
> Bear't that the opposed may beware of thee.*"
> — IBID.

> "*Thy head is as full of quarrels as an egg is full of meat.*"
> — IBID.

> "*I won't quarrel with my bread and butter.*"
> — SWIFT.

X.

GOVERNMENTAL ARBITRATION.

THE freedom of individual contract is the chief cornerstone in the structure of any system of liberal government. It is something that must be accorded and guaranteed to every citizen, whether he be high or low, rich or poor, employer or employee. Any legislation, or even prevailing custom, which tends to its impairment, is tyrannous. The greatest danger of the present time lurks in new forms of despotism imposed in the guise of humanity and philanthropy. The laboring man has more to apprehend from special legislation — ostensibly in his behalf — than any one else. However plausible new legislative departures may seem to him in their inception, their ultimate working produces hardship. His real interest demands free conditions, prevailing confidence, and general prosperity. To create demand for labor, there must be some inducement for starting new enterprises, and the extension of those already founded. Men of means will not embark in business with the prospect before them of interminable friction, or if the State, in response to demagogic demand, proposes to take control of their business and deny them the right of free contract.

During the last few years, many States have made provision for tribunals of arbitration, whose business is the settlement of disputes and controversies between employers and employees. These provisions for the machinery of arbitration vary somewhat in detail, but are similar in general plan and operation. General experience up to this time confirms the conclusion that no practical or permanent benefit can be expected from legal arbitration as a State system. It may be of some use, morally, as a temporary expedient to bridge over chasms of active hostility, or for emergencies when reason has lost its sway; but it is useless as a means for the permanent settlement of differences continually arising between capital and labor, while they occupy their present artificial and antagonistic attitude. Courts are already organized, and laws in force, to construe and enforce existing contracts; but the province assumed by these tribunals (at least in some States) of making new contracts between citizens, and of fixing prices other than those established by supply and demand, is a novel and unwarranted advance in the direction of paternal government.

What manufacturer can possibly have any security in engaging in business, if he be debarred from the natural right and freedom of hiring labor at its market price, or at a rate offered by those who are willing to sell? It is evident that no person or corporation will permit the State to transact their business for them; and if the State insists upon so doing, then business must come to an end. As well have a State board to determine the natural or proper market price for potatoes, clothing, or dentistry. These are the *product* of labor; and if the value of labor is to be arbitrarily fixed by the State, the same logic requires its application to them. During the French Revolution there was an attempt to do this impossible thing. Even if this were a proper function for this court, it is evident that in order to arrive at intelligent decisions, it must adopt rules and methods of procedure like a court of equity; that is, it must call in witnesses on both sides, and make up a verdict on the weight of evidence. It is also plain that even if the State usurped the right to make arbitrary contracts and prices between citizens, regardless of natural or market values, no board could possibly judge intelligently of the great variety of occupations, conditions, and questions that would come before it. It might be able and honest; but in addition, it would be necessary for its members to be universal experts. No two cases would be alike. It would not be simply a question of law and principle, or right or wrong; but, rather, of materials, qualities, markets, credits, competition, expenses, and many other elements which would all have a bearing. Aside, then, from its strained and unnatural jurisdiction, it would be a physical and mental impossibility for any board to grapple with such a variety of problems as would come before it.

Voluntary arbitration is of value in its proper sphere, but the fixing of prices and forcing them upon an unwilling purchaser is coercive. In the interpretation and enforcement of *existing* contracts arbitration is often quicker, less expensive, and more satisfactory than the regular process of law; but its adoption must be voluntary on both sides. The time-honored method of settling disputes by each party choosing one who is familiar with the conditions, and they choosing the third, the three then acting together to make a just settlement, is a commendable way of adjusting differences without requiring the intervention of a State board.

Conciliation, however, is more useful than arbitration. There is an important difference between them. The former may be employed regardless of State law, and is always mutually voluntary. Often all that is necessary to settle serious disputes is the assistance of conciliators who possess the confidence and esteem of both parties. They also should have a thorough knowledge of all the details and peculiarities of the special business, such as would be impossible with any State board. By such means angry feelings and prejudices may often be subdued, and reason and good sense brought to the front. When, in a conciliatory spirit, those who differ can be brought to sit around the same table and reason together in a friendly way, differences rapidly disappear. This would not often be the result of formal arbitration, which has the character of a court of law, in the fact that each side is arrayed against the other. Arbitration, in the proper sense of the word, must proceed under statutory or judicial authority. Even when both parties enter into it voluntarily, they must relinquish their freedom to a great extent by consenting in advance to accept the award of the arbitrators, so as to enable it to be judicially enforced. This gives it essentially the character of a court of law, with all its incidental antagonism and bitterness. If it have not these features, it is in reality conciliation, and not arbitration.

As long as the present strained and opposing relations exist between capital and labor, disputes and controversies will be numerous and bitter. Any ostensible settlement of them by boards of arbitration will only be a brief truce, rather than a treaty of peace. Under the head of conciliation is included all that is voluntary, friendly, reasonable, and fair in its character; and its possibilities for usefulness are great. Arbitration which must take account of the legal, opposing, and two-sided phases of a question, is well-nigh valueless for permanent results. There is much of the combative element in human nature; and instead of stimulating it to greater activity, it should be counteracted and subdued by other qualities which are just as inherent in man's constitution. Only by such means can the different elements of society be united and harmonized.

ECONOMIC LEGISLATION AND ITS PROPER LIMITS.

"*There shall be, in England, seven half-penny loaves sold for a penny: the three-hooped pot shall have ten hoops; and I will make it felony to drink small beer.*"
KING HENRY VI.

"*O! it is excellent
To have a giant's strength; but it is tyrannous
To use it like a giant.*"
MEASURE FOR MEASURE.

"*Let us a little permit Nature to take her own way; she better understands her own affairs than we.*"
MONTAIGNE.

"*For where's the state beneath the firmament
That doth excel the bees for government?*"
DU BARTAS.

XI.

ECONOMIC LEGISLATION AND ITS PROPER LIMITS.

To what extent the State may properly interfere with the industrial freedom of its citizens is a difficult and many-sided question. We shall not attempt to answer it in detail, but rather indicate certain general principles deducible from Natural Law, as a guide to its solution. The goal to be reached is the greatest good for the greatest number; and natural principles are the finger-boards that point out the way.

It is obvious that as modern civilization becomes more complex, population denser, and inventions and improvements more numerous, the scope of legislation, especially municipal, widens. The modern city, in many respects, is a great copartnership. Some sentimentalists hail municipal drainage, water-works, lighting, and possible rapid transit as successive steps in socialism. They are, however, only the wise business methods of a great corporation. Circumstances make them expedient, but their purpose is not to absorb private interests but to render them aid. The municipality can economically supply the citizen with water and light without the least impairment of his personal rights or privileges. With the growth of cities and profusion of inventions, an increasing number of functions can be performed by public administration. This is especially true, where, in services like those before mentioned, unlimited private competition is not practicable. The supply of water and light in a municipality involves the use of the public streets, which makes it a local, natural monopoly. Therefore, if not owned by the city, it must at least be regulated by it.

But any enlargement of public functions within the limit of practical private management increases opportunities for official waste, political corruption, rings, and spoils, with all their incidental demoralization. If the average city alderman and councilman were thoroughly expert and also conscientious and incorruptible, the question would be considerably modified. Even within its present limits the public service in most cities is shamefully defective.

As fundamental principles we may conclude first, that the State should not interfere in any enterprise that may be as efficiently carried on by private control; and second, that it should leave all questions of prices, rates, wages, and hours, to the natural regulation of free and untrammelled conditions.

Under the first of these propositions, let us note a few of the disadvantages of governmental management as compared with that of individuals or private corporations.

Many examples will occur to the mind of an impartial inquirer, showing the superior excellence and frugality of private administration over municipal, State, or national. The advantage is apparent, and usually involves not only cost, but efficiency and thoroughness in management and execution. For example, the public buildings of the United States, built by governmental or political organizations, on an average have cost vastly more than if erected under private management. It does not follow that this difference is always the result of dishonesty or mismanagement. It is in the nature of things; or, in other words, in accordance with Natural Law. The more close and direct the connection between the investor and the investment, the greater will be the economy and efficiency; and the more indirect and remote from the contributor or tax-payer the expenditure, the greater will be the waste, mismanagement, and extravagance. It would seem that those persons who are advocating governmental management for railroads and telegraphs must be blind to these laws, and to the teaching of experience. Take a great railroad system, the successful management of which requires the highest grade of executive talent, put it under the control of politicians of the dominant party, and the result may be easily imagined. In proportion as the domain of State administration is widened, the amount of "spoils," already too large, is increased; and these would be fought over under such a plan by politicians after every election. Divorce politics from any industrial enterprise, and a long step is taken in the direction of doing business on business principles. In the face of these undeniable facts, is it not strange that intelligent men urge, with evident sincerity, that the incubus of national and political control be fastened upon many kinds of business now efficiently conducted by private and corporate administration? It is evident that demagogism is the real foundation of many efforts in this direction. It is expected, as a matter of course, that when a city hall, court house, State house, or custom house is built, the expense will be much greater, and the utility less, than would be the case if the same were done by private enterprise. Official methods are extravagant, and operations under them are so hampered by "red tape" that they lack directness and efficiency. Rings, combinations, and favoritism are incidental to all such transactions. The opportunity for these abuses is much greater under our democratic form of government than with the nations of the Old World, whose powers are more centralized. There the civil service is more a matter of business and less of politics, and the administration of affairs is not continually changing. The necessary sphere of such governmental action among us is limited to those enterprises which, from their public nature, are beyond private control. In general, the rule of *laisser faire* has been the policy of our government in the past, and under it we have greatly prospered. The threatening evil of the present time is excessive economic regulation.

The second department of detrimental legislation named consists in the efforts to fix prices and rates, which in the end must inevitably be regulated by the law of supply and demand. This means a conflict between legislative and natural law. It may at first appear that some legislation of this kind would be beneficial, especially as applied to railroads. Whether or not correctly, the courts have decided in favor of the legality of State and national regulation of the rates of freight and passenger service. As this decision must be accepted, the only question remaining is that of expediency. It is urged that railroads are public highways, and that they have special privileges granted by their charters; and for these reasons they should be subject to governmental control. Quite an extensive test of this policy was made, a few years since, by the enactment, in a few of the western States, of what were known as "granger laws." Experience has shown that these laws were not only useless, but an injury to the public. It was only another of the oft-repeated attempts to substitute the artificial for the natural. Without State interference, business policy and competition are each constantly forcing the rates for service towards the normal standard, or to such a point as is natural and fair. Take, for instance, the worst supposable case,— that of a road without any apparent competition, either by land or water. The popular estimate of such a road is that it is a perfect "monopoly," and that its policy and interest will naturally cause it to make a tariff of high rates. A more careful examination of the case will show that it is against the true financial policy of even such a road to establish its rates above a fair standard.

Normal rates attract, foster, and increase both business and profits. Such a road, to be profitable, must adopt a policy that will encourage the location of manufactures, the development of agriculture, and the thorough settlement of the tributary territory. Sagacious railroad managers are learning that a large business at normal rates is far more profitable than a restricted traffic under a high tariff. In other words, they cannot afford to fix rates above the normal any more than below it. It is no doubt true that the managers of many roads have not fully realized the application of this general law; but as both experience and observation are persistent teachers, the tendency is constantly in the direction of a normal standard. In numerous instances, roads have voluntarily reduced their rates, thereby realizing as a direct result an increase of business and profits. As equipments and appliances have become more perfect, natural rates have steadily declined, and will continue to do so, regardless of legislation. Every reduction brings a great and unexpected increase of business. The problem before every railroad manager is to find as nearly as possible the normal basis; for in the end, that is the most profitable. In proportion as tariffs are removed from it, either above or below, profits decrease. Artificial restrictions prevent the increase of competition and discourage the building of new roads, as some of the "granger States" found to their sorrow, after the adoption of their "cast-iron" regulations.

About fifty years ago, "assize laws" were enacted in New York and some other cities, regulating the price and weight of loaves of bread, based on the price of flour. After a trial, which was attended with much trouble and expense, in consequence of the necessity for numerous inspectors, the laws were repealed. Besides the saving of expense, it was found that natural competition between bakers was much more effectual.

The old usury laws furnish another notable example of attempts to fix artificial prices. They were more of an injury to the borrower than to the lender. As well regulate the height of the tides by statute. When the artificial comes squarely in conflict with the natural, the latter will, sooner or later, surely triumph.

A striking instance of misapplied legislation is seen in statutes, existing and proposed, to regulate the hours of labor. These have been advocated and urged by so-called labor reformers and by labor organizations. They have brought a very strong pressure to bear on legislators in favor of these measures. When we look beneath the surface, and see their real effect, we can only be surprised that working-men are so blind to their own interests. Time is the one thing that all share alike. Unlike nearly everything else, the poor have the same amount as the rich. It is, in fact, the capital of the laboring man. By Natural Law, he has his full time to dispose of as he may think best. But when he asks for an artificial law, which forcibly, under all circumstances, will deprive him of the use of a portion of his own productive power, as by an "eight-hour law," he diminishes by so much his available reserve and renders himself poorer. This is the real sum and substance of restrictive legislation regarding hours of labor, whenever applied to adults. How unfortunate that American citizens should be so blind to their own interests as to deliberately beg to have their liberty and capital taken from them! If legal enactments be needed to prevent men from selling their time when they wish, it would logically follow that the State should control their eating and drinking, and prescribe their wearing apparel. It is a reflection upon the intelligence of the masses of the people to suppose that we have any considerable number of adult citizens who are so ignorant that they cannot decide for themselves how much to work. Moreover, if their physical welfare were promoted by shorter hours the decision should be voluntary. If the eight-hour law prevailed here and not in European countries, our manufacturers could not compete with theirs in the markets of the world. More workmen, too, would be attracted to our shores in hopes of an easier time. And both these causes would force down wages.

Legislation in regard to the frequency of payment of wages is clearly superfluous, though perhaps harmless, except as a precedent. Every unnecessary enactment decreases the respect for law, and lowers the estimation of its justice and impartiality.

The general demand for the widening of legislative functions arises, doubtless, from a vague though baseless idea that, by some additional enactments, evils which really come from defective character can be corrected by the magic of legislation. For this reason, an effort is made to correct every petty grievance by additional law-making. Prosperity is expected through some legal panacea, instead of by economy and industry. The present time is prolific of those so-called political economists who advocate new and unique additions to our already overburdened code. Ignorance of the first principles of political science can only give rise to such visionary theories. The "reformers" assume that all employers are blindly selfish, and wish to lengthen hours and depress wages. On the contrary, it is for the interest of every employer to pay good wages and make as short hours as competition and the nature of the business will warrant. Only by such a course can he retain his best help and get the highest quality of production.

Business prospers in the absence of legal interference, except to simply provide for justice and freedom.

It is true that the complex arrangements of modern civilization require State intervention in some ways unnecessary under more primitive conditions. The factory legislation of England, and similar enactments in some of our States, are examples. An excess of liberty to some individuals may prove a tyranny to others. As the good of society is more important than the possible advantage of one of its fractional parts, the operations of the few must be restricted when they encroach upon the liberty of the many. In other words, the natural law of liberty, as applied to society, is higher than that pertaining to the individual; and while they are not in opposition, the lower is modified by it. Thus human law should indorse and supplement Natural Law by restricting private will when it conflicts with the will of society. This is compatible with the greatest average freedom for all. The primary obligation of the State is in the exercise of what are usually known as police powers. There are a variety of other functions more or less intimately connected with these duties of protection to person and property. We expect the State to enforce our contracts, regulate our sanitary conditions, prevent and punish frauds, abate nuisances, and ward off general evils so far as is possible.

Among the examples of factory legislation which seem wise and proper is State interference in behalf of children whose parents or guardians, through motives of cupidity, will not protect them from overwork. The same restrictions in regard to an adult would be superfluous, for he is responsible and supposed to be able to judge correctly as to what is best for himself. Besides, if he be restricted in hours, it might mean for him less food and clothing and a poorer home. Wholesome regulations relating to fire-escapes, sanitary inspection, foul air, the fencing of dangerous machinery and elevator wells, are proper and necessary. They encroach upon no man's liberty, and private enterprise cannot always be relied upon to regulate them. Individual cupidity and neglect must be controlled by public supervision. Personal will must be subservient to collective will. Individual freedom might lead to the location of a powder-mill or a glue factory in a thickly settled street, unless it were restrained by collective freedom. It is obviously within the province of the State to appoint boards of health, and sanitary inspectors, whose duties shall include the supression of contagious and epidemic diseases, and the protection of air and water from pollution and contagion. As it is impossible for individuals to be universal experts, it is also necessary to have government inspectors to test weights and measures, to detect adulterations in foods and chemicals, and also, in some cases, to brand those articles of commerce whose quality or quantity cannot be verified by ordinary observation. Organized government has the power to aid and supplement the wisdom of the individual, without in any way restricting his independence, or deadening the competitive and elastic forces of the business world. The boundary line between State intervention and individual enterprise must, to a certain extent, be determined by a wise expediency; but the great end to be sought is, that private enterprise and competition shall be left unhampered. Any unnecessary dependence on the government for objects obtainable by private efforts is so clearly a violation of Natural Law that bad results are sure to follow.

DEPENDENCE AND POVERTY.

"*Practise yourself, for Heaven's sake, in little things; and thence proceed to greater.*"
EPICTETUS.

"*Honor and shame from no condition rise;
Act well your part, there all the honor lies.*"
POPE.

"*The poor always ye have with you.*"
JOHN xii. 8.

"*Teach every man to spurn the rage of gain;
Teach him that states of native strength possess'd,
Though very poor, may still be very bless'd.*"
GOLDSMITH.

"*To a close-shorn sheep God gives wind by measure.*"
"*Help thyself and God will help thee.*"
HERBERT.

"*The poor must be wisely visited and liberally cared for, so that mendicity shall not be tempted into mendacity, nor want exasperated into crime.*"
ROBERT C. WINTHROP.

"*Let not thy mind run on that thou lackest as much as on what thou hast already.*"
MARCUS AURELIUS.

"*There is nothing either good or bad, but thinking makes it so.*"
SHAKESPEARE.

"*Blessed is he that considereth the poor.*"
PSALM xli. 1.

XII.

DEPENDENCE AND POVERTY.

THE terrible degradation, vice, and poverty which prevail, especially in the slums of our large cities, furnish a difficult social problem for solution. That the vast majority of the victims of these conditions is composed of the unassimilated alien element is well known. But racial differences alone do not account for the great contrast. The native citizen has a much greater inborn independence of character. He may be equally poor and uneducated, but he will put forth herculean efforts to avoid falling into a condition of dependency. For generations he has been self-reliant and self-respecting. On the contrary, the traditional paternalism of European monarchies is reflected in the character of the immigrant. He has looked upon his government as an institution to lean upon, and it is difficult for him to grasp the spirit of Americanism. Thus a certain proportion of foreign immigrants, without much resistance on their own part, settle down into low physical, mental, and moral conditions. Their specific gravity is weak.

While no graphic descriptions of the deplorable conditions existing in our very midst can equal the reality, there is a deeper question involved, as to the effect and utility of holding such delineations continually in the public gaze. Many well-meaning and conscientious philanthropists have given us volumes of appalling detail of the "social cellar," hoping thereby to arouse public sentiment so that some radical and effective remedial measures would be inaugurated. Lacking, as they do, a full recognition of *law*, these writers work in the illusive light of a mistaken mental philosophy. Instead of *arousing* the public mind they harden it. The human mentality is so constituted that the iteration and reiteration of these "tales of woe" finally cause the things depicted to be taken as a matter of course. Instead of being regarded as abnormal conditions to be cured, they come to be accepted as inevitable and hopeless. The real, though by no means the intended influence of these graphic pictures of the nether side of humanity, is pessimistic and discouraging. To magnify abuses until they seem to be the rule, is to promote them. In varying form the old question of realism *versus* idealism is ever cropping out. Shall the worst or the best be made of existing conditions? We take direct issue with the host of noble men and women who, with the best of intentions and with artistic skill, have drawn, framed, and hung up vivid pictures of human misery. Pessimistic realism contains no element of cure. Such has been the conventional and largely the ecclesiastical way of dealing with evil, moral and social, for centuries, and little advance has been made. It has been thought that to hold up the abnormal, turn it about in the light and analyze it, would effectually make it detestable. It has rather made it familiar and expected. The invisible powers of thought and suggestion, of which the great majority are unconscious, are active in sociology as elsewhere. It may be thought to be outside the scope of this study, but it is nevertheless true that by psychological law, conditions which are widely lodged and spread out in the public mind tend to externally actualize themselves. There is almost such a process as thinking things into existence, and this most of all when they are not wanted. The more the animalism of mankind is put on exhibition, the more its corresponding unisons vibrate within. The same principle is seen in politics. The general suspicion and assumption of motives of self-seeking which are focused upon every one in political life is appalling. No difference how conscientious a man may be, a deluge of low insinuations is poured upon him. Is patriotism dead? If not, it survives in spite of those who see the worst in every man, thing, and system.

You would not wish your family to frequent the slums in real life, and if so, it is better not to do it in books. Pope's familiar lines regarding "vice" contain the germs of a whole system of philosophy. Shall all, then, avoid the slums for fear of contamination? Exactly the reverse in the case of every one who will go with any *intelligent* aid. Let those go who are able to discover there a little that is good for a foundation. There is a trace of wholesomeness everywhere; and where there is a minimum, it is of the highest importance that that little be recognized, stimulated, and made the basis of more. The slums are already dark with pessimism, and require nothing so much as a flood of optimism. They need cheer and hope. Condolence and sentimental sympathy so profusely offered should be replaced by encouragement. Emphasize the one talent they have, and this will bring five or ten into manifestation.

The pessimist has eyes only for the worst, and his presence is a black pall. The scientific way to cope with negative conditions is to displace them with positives. To hold up before any one his degradation, poverty, or ignorance, is to impress them more deeply upon him. The inhabitants of the slums need hope as much as food or fuel; in fact, the very lack of it is mainly responsible for their present condition. Every unfortunate has at least some small solid spot to build from.

There are still a few healthful germs in the darkness of the social cellar, and sunshine, air, and cultivation will make them sprout and grow. As the oak lies enwrapped in the acorn, so ideal possibilities are latent in all sorts and conditions of men. In the light of evolution, low grades are only such relatively.

In proportion as mind becomes pure and wholesome, habitations and environment are transformed as a resultant correspondence. Moral, social, economic, and hygienic education in individual character is the urgent necessity. If, without these, the population of the slums of New York were moved into the palaces of Fifth Avenue, the improvement would be more superficial than real. Model houses for the poor are good, but their faultless qualities can be maintained only from within. The charity demanded by the times is not so much an infusion of dollars as of moral and industrial training, together with the inspiration of hope, ambition, and independence. This necessity is becoming widely recognized; and university settlements, industrial, technical, and cooking schools, kindergartens and missions, are being multiplied. While pessimism paralyzes, optimism gives new life. Charity is being more intelligently administered and becoming more true to the name. The almsgiving of the past has largely been contrary to Natural Law and often worse than useless.

Turning from the slums to the broader applications of Charity, the great need is a scientific, rather than a sentimental basis. The laws of dependence are as exact as those of chemistry. The charitable societies of London are far more numerous and wealthy than those of any other city, and no where else is there such a vast amount of abject and hopeless poverty. What is the relation of these two facts; or, in other words, which is the cause and which effect? If we study human nature in the light of Natural Law for the solution of this problem, and also observe carefully the teaching of experience, we find that supply and demand equal each other here, as in the domain of commerce. Let the supplies of charity be doubled or quadrupled, and the demand from dependence keeps pace with them. These relations and sequences, being uniform, prove that they are not a matter of chance, but rather are governed by natural and unvarying principles. As rapidly as dependence can find something to depend upon, it will increase. In contrast with London is Paris, where racial conditions and customs would lead us to expect more and worse poverty. There is instead much less, and of a milder variety. The French capital makes but a moderate showing in charitable organizations when compared with London, where the number of old and thoroughly equipped benevolent associations is remarkable. These illustrations, and similar ones which might be cited, do not prove that charity is an evil. It is *misapplied* charity, which is really not charity at all, of which the world has been full, that is out of harmony with natural principles.

Charity is divine, heaven-born, the brightest and noblest of all virtues; but this does not alter the fact that so-called charity, misapplied, breeds dependence with unerring certainty.

The diseased, aged, and helpless are within its sphere, and he who has surplus wealth gets real sweetness out of it by applying it to lessen the misery and lighten the burdens of this ever-present class. Natural Law is not uncharitable nor mechanical, as some might hastily conclude; but compassionate and bountiful, when not transgressed and defied. Benevolence is normal, and the hospitals, asylums, and other humane institutions are entitled not only to our merciful regard, but we owe them a debt. Charity is a natural quality, and it would be unnatural not to exercise it. It is, perhaps, fortunate for society that it has its helpless and dependent class, for it furnishes an ample field for the exercise of the kindly and brotherly motives of man's nature. While all these facts cannot be too greatly emphasized, it remains true that every man who has in him the possibilities of independence, is degraded by opportunities to lean upon anything outside of himself. The contrast is the widest possible between the results of charity exercised in its true sphere, and those of its abuse, or when applied outside of its legitimate functions.

The so-called paternal governments of Europe have in them elements which tend directly to add to the numbers and degradation of the dependent classes, and to make their condition more hopeless and fixed in its character. It is just as demoralizing and destructive to a self-reliant manhood to lean upon the State, as upon some private organization. A government that upholds the rule of *laisser faire*, or non-interference, is that under which true manhood and independence are best developed and cultivated.

Mrs. James T. Fields, in her admirable book, "How to help the Poor," says: —

"To teach the poor how to use even the small share of goods and talents intrusted to them proves to be almost the only true help of a worldly sort which it is possible to give them. Other gifts, through the long ages tried and found wanting, we must have done with. Nearly a million of dollars, in public and private charities, have been given away in one year in Boston alone; and this large sum has brought, by way of return, a more fixed body of persons who live upon the expectation of public assistance, and whose degradation becomes daily deeper. The truth has been made clear to us that expenditure of money and goods alone does not alleviate poverty."

A sharp line of demarcation needs to be drawn between a poor man and a pauper. There is little necessary resemblance between poverty and pauperism. The worst calamity that can befall a poor man is to become pauperized. He who blindly scatters money in the name of charity is liable to do incalculable harm. On the other hand, he who teaches one how to help himself, and raises him from the dependent class into that which is thrifty, does society and humanity a great favor. No person of means can discharge his obligations to society by careless and indiscriminate giving. Industrial schools, and any other aids that teach the way of self-support, and give the young such a training as will put them on their feet, deserve the most liberal support and encouragement. Help some dependant to discover a path of self-support; for by this act of real charity you bring him into harmony with Natural Law, and no gift of money could equal that favor. The knowledge of something to fall back upon in the future, outside of one's own exertion, causes improvidence in the present. The tramp who knows that charity and the soup-house are in readiness for him when winter comes, will not put forth much effort to find employment during the summer and autumn.

It is not within the province of this book to present statistics to prove how much the dependent and pauperized classes are increased by intemperance, vice, and crime. That these, however, are the true causes of nine-tenths of the poverty, misery, and degradation is evident to any candid observer. It is idle and fallacious to attribute evils due to these causes to any inherent fault of our social system.

SOCIALISM AS A POLITICAL SYSTEM.

"*Knowledge is the only fountain both of the love and the principles of human liberty.*"

WEBSTER.

"*Where law ends, tyranny begins.*"

WM. PITT.

"*Of what avail the plough and sail,
Or land or life, if freedom fail?*"

EMERSON.

"*O Liberty! Liberty! how many crimes are committed in thy name!*"

MADAME ROLAND.

"*Man is not the creature of circumstances. Circumstances are the creatures of men.*"

DISRAELI.

"*Sail on, O Ship of State!
Sail on, O Union, strong and great!
Humanity with all its fears,
With all the hopes of future years,
Is hanging breathless on thy fate!*"

THE BUILDING OF THE SHIP.

XIII.

SOCIALISM AS A POLITICAL SYSTEM.

THERE are few terms in common use so elastic in definition, and which signify so many different things to different minds, as socialism. In its root meaning and derivation, it is both harmless and attractive. To be social is creditable. It carries the idea of friendliness, companionship, brotherhood, neighborly interest, and even unselfishness. These are some of the ideal qualities of humanity, and it is impossible to over-estimate their importance and beauty. If socialism promised their general embodiment, it would be supremely desirable and could not come too soon. There are some philanthropists who call themselves "socialistic," and persuade themselves that, in some way, the government can take hold of the matter and speedily usher in the reign of these delightful conditions. There are sentimental clergymen, who look upon business from the outside, and know little of its inherent self-regulative and compensatory laws, who wish to see a New Order, from the fact that they are unable to discriminate between the present natural *system* and its *abuses*. They forget that everything normal has its negatives and violations. There are "Christian Socialists" who are unmindful that He whom they regard as their perfect and complete Model labored entirely within the domain of human life and character, and not at all in external political affairs, which, in their very nature, are only expressive and resultant.

Some enthusiastic theorists hail every little widening of state or municipal functions — made necessary by advancing civilization — as "socialistic," and harbor the pleasant illusion that society is shortly to be reconstructed by a short-cut process. But the beautiful social qualities before enumerated are not political, but moral and personal. Their location is in the individual, and their exercise comes through voluntary growth and unfoldment.

Yielding to no one in our admiration for socialistic qualities in moral character, *socialism* is here considered as a political system. This is its claim, and there is no other logical method. The accepted definition of the term involves the fundamental political reconstruction of society. Any such radical change must necessarily be coercive and not evolutionary. Socialism, as a system, means not merely a friendly interest in our neighbor's welfare, but a formal and forcible one. Instead of natural liberty, it signifies artificial interference, even though imposed in the name of brotherhood. It would ignore inherent, elastic, self-regulative forces, which are omnipresent, and prescribe everything by mechanical metes and bounds. Ignoring spontaneous individual growth, it would furnish universal moulds casting all shapes in stiff and arbitrary form.

Socialism, as a possible political framework, is not only fatal to all evolutionary social development, but is paralyzing to all ideal human brotherhood. If it were possible to make men altruistic by legislation, all its sweetness would vanish with the loss of its voluntary and spontaneous spirit. But legislation piled Ossa on Pelion will not change human character. As well galvanize a decaying body into youthful vigor as to inspire brotherly love, or even morality, by coercive legalism. As properly call a stick of timber a tree, as to denominate political socialism brotherhood. A "whited sepulchre" is a sepulchre still. Were it possible to put in motion an entire paraphernalia of outward balances, checks, weights, and measures, living benevolence and unselfish service would become extinct.

A non-recognition of the self-regulative and educational forces of Natural Law, everywhere present, leads many to conclude that there is but one way to get rid of injustice, and that is to pass legal enactments against it. It is like building a dam across a stream. As the water behind it rises they would build it higher, and patch it here and brace it there. But if it be not swept away, the current, in full volume, soon flows over it. Mistaking abnormities, which come from human distortion, for normality, they would raise an artificial wall to forever stop the rising flood of evils.

Legislation is invoked to cure all the ills "that flesh is heir to." Is there injustice? let us make a law against it. Are hours too long? shorten them by law. Is there too much competition? put it down by law. Are there trusts? wipe them out by law. Are times hard? improve them by law. Is money tight? make more by law. Is the millennium slow in coming? invoke the law. An able writer, in a recent article, characterizes the proposed phenomenal legislation as the modern tower of Babel. He says:—

"And now there comes a band of earnest men and women who see plainly the evils of the times, and who would give up much to help their fellow-men. 'Come brothers,' they say, 'let us be brothers indeed. We will make a tremendous, a sky-reaching, an all-powerful law, that all men are and shall be brothers; that no one shall have more of this world's goods than another; that each shall give his best work and his best endeavors for the common good of all. We will all work for the Government, and the Government will feed us all. We will have no more poverty and no more riches, but all shall work and eat at the nation's table, and none shall be kept back in idleness or go away in hunger.' This is the plan of the Nationalist. It is the loftiest structure of its kind that the mind of man ever sought to rear; for socialism thinks to outwit Mother Nature herself, and to legislate the law of the survival of the fittest off the face of the earth. It is the modern tower of Babel. But it is not to be built of bricks, but of men; and the mortar of legislation never can make a man stay put. The law of evolution is superior to the laws of men. Before man was, it was, and yet, like the Babel builders of old, he thinks to overtop it."

As a moving body seeks the line of least resistance, the average man will make a living with as little exertion as possible. Nine times out of ten, as has often been demonstrated, if he were in the man's shoes whom he regards as an "oppressor," he would be the harder man of the two. What every live man needs is not more law but less. He wants natural freedom, consistent with the freedom of others, and has no use for arbitrary trammels.

The two wings of the socialistic propaganda are very unlike. At one extreme is a small band of earnest souls, sincere and benevolent, though impractical. They are filled with a fraternal spirit themselves and wish every one else to be. But under the same banner, though ten times as numerous, are those to whom socialism means, not more fraternity, but a grand divide. Avarice and envy are covert elements in human character which sway men powerfully, even though often unconsciously. The vast majority of socialists whose ranks shade through different degrees into red anarchism, gather encouragement and strength from the little section of sentimentalists who comprise the wing that is in sight and does the theorizing. The crowded socialistic columns that loom up in the dark background are looking forward to the time, when, through the forms of law, the estates of the more thrifty may be confiscated, which they imagine will give them governmental support and an easy time.

It is idle to claim that the ignorant and unassimilated alien elements which form the great bulk of the socialistic party of America are actuated by a fraternal or unselfish spirit. Their lurid declamations against property and capital, and the spirit, even of their best literature, are conclusive on this point. Their prevailing animus is distinctly destructive and not constructive. When the French Revolution was kindled, the theory was "Liberty, Equality, and Fraternity." In practice, it turned out to be, tyranny, cruelty, and destruction.

Were it possible to inaugurate political socialism even without violence, it would smother personal ambition and liberty, and discourage progress. As is now the case with penitentiary labor, the minimum would speedily become the maximum. As a practical result, consumption would soon overtake and pass production, including the reserves of previous accumulation. Then would follow famine, civil war, and anarchy, and the whole artificial conglomeration would fall to pieces from its own confusion and corruption. Brute force and chaos would prevail, until, at last, the few survivors would have again to make a new beginning on natural principles. But complacent theorists say, "we shall accomplish it through evolution without revolution." But if it is a good and desirable condition, the sooner it comes the better, and such is the reasoning of their impatient followers. In this way gigantic but irresponsible forces are directly stimulated.

Socialistic agitators descant upon "wage-slavery," but that is nothing compared with a coercion which would sweep away all liberty. The employer — and the vast majority of employers are not rich — is a "slave" to the markets, as much as the wage-earner is to his toil, and often more, for he cannot so easily change his position. It would indeed be slavery to have eating, sleeping, clothing, working, and all the social and personal activities conducted upon the compulsory plan, in which each is assigned his place by the "majority," which would really consist of a few official dictators. The blotting out of individual liberty would mean *real* slavery. There would be no incentive for personal effort, such as is now afforded by the hope of providing for infirmity or old age, or for the wants of family and kindred. The fruits of a man's industry would belong to "The State." The choice of occupation would be dictated by the office-holders of the dominant party.

The mild socialistic theorist is like a man who is unwittingly playing with fire in the midst of a wilderness of inflammable material. While he is innocently dreaming of coming fraternity he is unconsciously though forcibly appealing to some of the strongest and lowest passions in mankind. These propensities are wholly on the material or animal plane. Political and coercive socialism is thoroughly materialistic. It minimizes character and manhood, and magnifies the value of externals. When the theoretical socialist indulges in exaggeration about the unequal distribution of wealth, *as attributable to the present order*, he directly appeals to the envy and avarice of the ignorant and selfish. They are made to believe that the reason why they have not as much wealth as some one else is because they have not had their *rights*. Socialism is no question of the poor against the rich. It would be as disastrous to the former as to the latter. It is a question of thrift, industry, economy, and character, against dependence, shiftlessness, and avarice. It would prostrate individuality instead of placing it upon its feet.

There are marked inequalities among men and elsewhere, and there will be until the law of evolution is repealed and the universe reconstructed. But the divine order provides that those who are most advanced shall be an aid and inspiration to those who are in the rear. The spirit of a voluntary and altruistic brotherhood was never before so active, and nothing but a cold and selfish legalism can chill it. The overflow of charities, aids, and helps is rising in volume. Men are recognizing a racial unity, even though it be made up of diversity. Voluntary virtue grows and glows in profusion and purity, but a coercive element introduces formal legalism and coldness. There always have been, and will be, leaders and followers in knowledge, power, wealth, invention, science, and philosophy, and the world would be in a sad condition if all were levelled down to a dead common-place. The "fittest" portray the possibilities and inspire the ambition of the less fit. Such is the law of growth which always progresses from within. The hostility that is being worked up against honorable accumulation, upon the theory that "property is robbery," is an outcropping of barbarism and an invitation to return to it. The proposed legislation of some of the trans-Mississippi States would lead one to suppose that capital is not only not to be invited, but something to be kept out.

The present labor-value product of a manual workman, as measured in comfort and luxuries, is about three times what it was forty years ago. The advance has been natural and healthful, and will be continuous under normal conditions. This represents the advance of a class, but in America individuals are not bound to a class. Those who are self-fitted leave it behind. It is preposterous to make "the social system" the pack-horse for the huge load of negations, sins, and weaknesses that inhere in ignorance and imbecility.

Through the rosy vision of the theorist, "The State" — which is the all-comprehensive agency in socialism — will be a perfect, omnipresent, and omnipotent instrumentality, able not only to cognize every detail, but to control universal equity and righteousness. But the real State would be composed of office-holding politicians. With tenfold greater opportunities than present conditions afford, the probable reign of dictation, jobbery, and favoritism may be faintly imagined.

It is foreign to our purpose to attempt any historical or detailed study of voluntary local socialistic experiments which have been made. Though differing somewhat in doctrine, they have positive features in common, which are visionary and abnormal.

At one period in his career, Horace Greeley gave local voluntary communism an earnest investigation. His opinion was formed after close and practical observation. He says: —

"Along with many noble and lofty souls, whose impulses are purely philanthropic, and who are willing to labor and suffer reproach for any cause that promises to benefit mankind, there throng scores of whom the world is quite worthy, — the conceited, the crochety, the selfish, the headstrong, the pugnacious, the unappreciated, the played-out, the idle, and the good-for-nothing generally; who, finding themselves utterly out of place and at a discount in the world as it is, rashly conclude that they are exactly fitted for the world as it ought to be."

Socialism is not indigenous to American soil, and is an exotic in any country where free and constitutional government prevails, though it assumes to oppose despotism. As by Natural Law extremes meet, so violent or compulsory socialism becomes itself despotism. Its apostles and advocates are not numerous among native American citizens, or even among those of foreign birth who have any intelligent appreciation of our political system. These irreconcilable extremists are willing to ingulf society, themselves included, in general ruin, and to relapse into consequent barbarism, rather than that existing civilization and government should continue. That they gather moral encouragement from milder socialists, some of whom advocate the same end, but hope to bring it about by peaceful means, is beyond a doubt. The pronounced sentimentalism of the times, which is making such efforts to set aside natural principles, is, though perhaps unwittingly, lending encouragement in the same direction. The warfare against Natural Law is carried on by an army of allies whose several motives and aims greatly vary, but in this general hostility they are a unit.

Experience, as before noted, which is the indorser of

law, shows the uniform failure of socialism even in small select communities under conditions highly favorable to success. Voluntary socialism, under the most flattering circumstances, and with the most conscientious and enthusiastic leaders, has been experimented with again and again. It is true that in certain instances, societies having socialistic features have existed for a while, but in none has there been vitality and growth. From "Brook Farm" down to the present time, there have been occasional bands of impracticables who have repeated the experiment. Of course such little local voluntary communities are perfectly harmless and have no likeness to a general political system. They, however, furnish a test of the principles under a thousand-fold greater chances for success than would be possible for the proposed New Order. If such mild and promising examples have proved futile, what might be expected as the result of a violent and compulsory commune, attempted, not with a voluntary and picked community, but with all the heterogeneous elements of society? A menagerie let loose would be a fit illustration of the result. If attempted, it would very likely produce an upheaval similar to the French Revolution. World-wide experience, as well as the teachings of Natural Law, prove the truth of the proposition, — *that the condition of civilization or barbarism among nations is in proportion to the security and inviolability of individual property rights.* Adam Smith asserted that the security afforded to property by the laws of England had more than counterbalanced the repeated faults and blunders of the government. It is not too much to claim that the foremost and commanding position of England and the United States of America among the nations, is due to the safeguards erected around property rights, and the but slightly obstructed operation of natural laws by governmental or other interference. No nation can be named where private accumulations are insecure, in which there is not a coexisting state of barbarism. These truths are so obvious that it seems superfluous to demonstrate them. But the fact remains that charlatans in political economy are making great efforts to disseminate opposite theories, and apparently with much success.

It is the main province of legislation and political science to provide the best and surest means for protecting industry. This is all-important, for the reason that the right of personal accumulation is the most powerful of all encouragements to energy, thrift, and the increase of wealth. *The certainty that a man can enjoy the fruits of his toil is the great stimulus to production, enterprise, and prosperity, with the individual and with the nation.* In those parts of the world where the title to property depends upon a strong arm, or where it is liable to confiscation by the ruling power, production is confined to its rudest and most primitive forms. The doctrine of general or governmental ownership of land — which some visionary but well-meaning people think would "abolish poverty" — is already in force in large sections of Asia and Africa; and as a natural result, there is no fixed property except of the rudest description, and valuables are either hid in the earth, or quickly carried by caravans to places where private ownership is recognized and protected.

With human nature as it is, how many would be provident, industrious, or economical under the most perfect system of socialism yet conceived? Enterprise, ambition, invention, and progress would all wither, as if under the shade of the deadly upas. If an ideal millennium had come upon the earth, so that men loved others more than themselves, there would be true moral socialism from within; but until such a time, civil law and government will be indispensable.

The genius of socialism seems to be embodied in the old adage that "the world owes every man a living." No matter how lazy, improvident, or reckless he may be, his industrious neighbor, who by patient toil has become the owner of accumulated labor, is expected to divide with him, and, in future, to keep on dividing.

Socialistic agitators ring so many changes on such recently coined phrases as "impending revolution," "wage-system slavery," "industrial crisis," etc., indicating some expected revolution, that some persons actually look for a time not far distant when the government, through wholesale confiscation, will be able to take care of them, and work be a thing of the past. Is it a wonder that great masses of ignorant immigrants become saturated with such ideas, when it is considered that socialistic, atheistic, and anarchic literature forms their chief intellectual diet? Many of them remain in solid, unassimilated masses, and learn little or nothing of our institutions or language. Here is a fertile field in which to sow the seed of moral and economic truth. The right sort of reading matter in their own tongues would do much to neutralize the baneful influences which loom up on our national horizon like a black cloud.

CAN CAPITAL AND LABOR BE HARMONIZED?

"*One touch of nature makes the whole world kin.*"
TROILUS AND CRESSIDA.

"*He had a face like a benediction.*"
DON QUIXOTE.

"*Not chaos-like together crush'd and bruis'd,*
But, as the world, harmoniously confus'd,
Where order in variety we see,
And where, though all things differ, all agree."
POPE.

"*If thou shouldst lay up even a little upon a little, and shouldst do this often, soon would even this become great.*"
HESIOD.

"*Gain not base gains; base gains are the same as losses.*"
IBID.

"*Brother, Brother! we are both in the wrong.*"
GAY.

XIV.

CAN CAPITAL AND LABOR BE HARMONIZED?

CAPITAL and labor, being natural interdependent conditions, are already harmonious. It is only personalities that are discordant. The conventional arbitrary division of society into two parts, respectively termed capital and labor, is prejudicial and misleading. Labor is like a tree of which capital is the fruit. The sentimental antagonism between the two which has sprung up in many minds — and it has no other existence — is unfortunate for both. Capitalists and laborers are relatively good or otherwise, but capital and labor are only good. How can two parts of a unit, each utterly incomplete without the other, be naturally antagonistic?

Human activity systematically applied, is labor; and the outcome, whether large or small, is product, or capital. Capital is only a name for preserved or stored-up labor. A stock of bows and arrows, baskets, or skins is the capital of an Indian, and the preserved harvest that of a farmer. Capital may be visible or invisible, material or mental. The wage-worker, and no less the school-teacher, and even the pupil, are all laboring to produce it.

A baseless and wide-spread fallacy exists, to the effect, that, in some way, there is a limited and fixed amount of capital in the aggregate, so that if some have more, others must have less on that account. On the contrary, capital is like seed; it tends to propagate itself, though not necessarily by making its original owner poorer. There is a constant *creation* of capital, and the larger the amount already in existence, the more easily and rapidly additions are made. There is a popular feeling that the very wealthy are "monopolists," and even their inanimate possessions almost seem to come in for a share of opprobrium. But however unwise, or even selfish, persons may be, product is good. It has both intrinsic and representative value.

The Astor estates in New York are examples of great accumulation. But in each and every building belonging to them labor, skilled and unskilled, has been employed, paid for, and stored up in every detail, from foundation to cap-stone. Their care and repairs will also require future labor indefinitely. As wealth accumulates, it calls for higher and finer grades of production. Primitive communities have little use for artists, carvers, decorators, sculptors, and frescoers. This is no apology for wealth, for none is required. It is only a study of the laws, demands, and methods of an opulent civilization. If half a dozen wealthy proprietors locate together, even in a wilderness, labor is stimulated and growth takes place on every side. Stored-up labor not only pays taxes, but constantly demands *active* labor. Even its net income is reinvested, and goes to swell the great current of business enterprise. In those countries or cities where the economic accumulations are great, the poorest inhabitant shares their advantages. The public parks, libraries, and art museums which are accessible to every one in all large cities, exist only as the fruit of great and concentrated wealth. A public garden or flower-lined boulevard, made possible only by a vast surplus of stored-up labor, exhales its beauty and fragrance to the penniless visitor as much as though he were its sole owner. The modern municipality consists of a boundless amount of human energy and skill in preservation. To be consistent those pessimists who count "property as robbery" should remove to a desert.

Labor-value is fixed by the average opinion of mankind,

and depends upon the intelligence back of it as displayed in the quality of product. Any other appraisement must necessarily be purely artificial. The world will freely pay a hundred times as much for a painting by a Meissonier as for one by an ordinary artist, and no theorizing will change such a relative valuation. Is this fact a hardship to the latter? No; because his production is no worse on account of the existence of a Meissonier's, while, on the other hand, he is furnished with an ideal which will more and more inspire every stroke of his brush.

Capitalists and laborers mingle in all degrees. Many highly successful employers work more hours than their employees; and, as is well known, care and responsibility are often more wearing than physical toil.

Under normal conditions, in the absence of strikes and obstruction, the demand for labor by advancing civilization will always equal the supply at steadily increasing rates. This will be no less true if labor-saving machines are yet multiplied, and even if new motors are discovered in addition to those already utilized. Efforts, however, to force labor-values beyond the smooth working of Natural Law destroy confidence and react upon and injure the very cause they are expected to aid. While wages have steadily increased, the value of capital, which is measured by the rate of interest, has materially declined.

Occupation is indispensable to human progress and harmony. The drone, whether rich or poor, who lives in idleness or for selfish gratification, reaps the inherent penalty, which is decay. A large part of the discontent among manual laborers arises from the mistaken idea that happiness and contentment naturally come from wealth. But man is so constituted that he can absorb only a certain limited amount of material good, and every attempt to do more results in satiety and disappointment. Most men refuse to learn this lesson, except as the bitter result of experience, but it is a fact that the sons of rich men begin active life heavily handicapped. Human life is barren and disappointing unless inspired by an abiding and worthy *purpose*, and no talent grows except through faithful exercise. Contributions to the world's wealth are as genuine when made in knowledge, science, art, or research, as in food or clothing.

The claim is made on the part of labor that it does not receive a fair share of the profits of production. What constitutes a just division of these products? Exactly what the so-called labor reformers and sentimentalists demand is a matter of great uncertainty. The only unanimity among them is in dissatisfaction. As any proposed new division must be made by ever shifting artificial rules, there could be no substantial agreement. When the solid ground of natural principles is abandoned, the restless currents of sentimentalism reveal no resting-place.

It is not strange that manual laborers often feel dissatisfied. As a rule, they toil hard for a very moderate subsistence. When they look around and see many who have a surplus, they think there must be something wrong in a system under which there is such inequality. But such reasoning is superficial. Men are created with unequal capacities and powers, and it is beyond human ability to equalize them. Society could as effectually resolve that two and two make five. The world's conclusions cannot be arbitrarily set aside. It values mental force at a higher rate than manual, and it would be as futile to attempt to change these conclusions as to level the Alps. Society is exact and unerring in its estimates. It marks its valuations on both mental and manual force with as great a degree of accuracy as is seen in the coinage of a mint.

The brain force of a McCormick, which conceived the reaping machine, was greater in the results of its production than a million strong right arms, each of which could wield the sickle. The world, therefore, makes its appraisement of his product at millions of dollars, and willingly pays the obligation. Whether or not we like it, this law cannot be repealed. The brain power, not only of inventors, but of all those who possess the ability to organize and execute, is scarce, sought for, and therefore has a high valuation. The mental force that organizes, builds, and puts into operation a great railway system is worth, perhaps, millions,[1] because its product may be the settlement and development of two or three States or Territories. If this kind of force were more plentiful, the world would not put such an extravagant valuation upon it. A thousand muscular bodies may be found as often as a single brain of this quality. No amount of sentiment can change the arrangement of these evolutionary principles. Were it in our power to explore deeply enough, we should probably find that it is even best as it is. It is only the few who are skilful in originating enterprises, and in conducting them to a successful termination. They also have a better knowledge of Natural Law, which they make the most of by securing its aid. If the many could command all these advantages of mental power, there would be a much wider table-land of equality. It is now only the lofty peaks of attainment and production that attract special attention. Having found that inequality is universal and based on law, it is unwise to complain of it, and foolish to expect to abolish it. Did the Creator make a mistake when he instituted the evolutionary order which makes men of unequal capacity? As it is, every man gets the reward which comes from the exercise of his own productive energy.

[1] We do not forget that there are men of this class who have amassed large fortunes by stock manipulations which are illegitimate. Such an exercise of mental energy is unnatural, perverted, and hostile to the best interests of society. Natural Law would sanction restrictive legislation when applied to such artificial operations, and also the positive and sure punishment of every form of dishonesty.

This fact furnishes a continual stimulus to the lower to advance towards the higher. Were it not for brain labor, we should still be in barbarism. It is the increased production of the mental force of the few that has developed civilization. Labor, which is now making complaint, is getting a large share of the benefits of this improvement, its blessings being enjoyed even by the humblest.

Capital is only the surplus that is saved above consumption, and it is not only the progenitor of civilization, but it is all that gives value to labor. Without it there would be no demand for labor. So far, then, from being envious of another's greater attainment, we should rejoice over it; for we are better off than we should be otherwise. The capitalist who, with executive talent and millions of money, has built a railroad, has done a great favor to labor and society. Boundless acres, before useless, are by its influence transformed into fruitful farms. Thousands of laborers, in addition to those who receive direct employment from its operation, thus find sustenance and occupation. It is a fallacy that the presence of the very rich in society tends to make the masses poorer. It is exactly the opposite. The sentimental and false ideas now prevailing on this subject are the fruits of demagogism and envy. There is a kind of discontent which is wholesome, for it stimulates effort; but the variety now prevailing seems to be of the envious kind, for its spirit is to pull down rather than build up. If these conclusions are correct, it follows that improvement for wage-earners must be looked for in harmony with them. We shall succeed if we call to our aid the powerful machinery of natural principles, but fail if we challenge and defy them. There is no panacea nor charm by which poverty may be abolished, and no magical cure for the ills of society and inequalities of fortune. There is, however, room for vast improvement, if sought in the

CAN CAPITAL AND LABOR BE HARMONIZED? 165

right direction. We must work along the lines of Natural Law, instead of trying to cross them at right angles.

Before indicating more fully how the relations between capital and labor may be harmonized, let us note briefly some things which cannot accomplish it.

It cannot be done by combinations of like elements, as of laborers with laborers and employers with employers. Natural competition always exists between occupations which are alike. If, therefore, a number of carpenters organize an artificial combination which holds them together, it is in direct opposition to the law of natural competition. It is a combination all on one side, and is as incomplete as a four-wheeled carriage would be with two of its wheels removed. These carpenters are helpless because there is no demand but only supply in their combination. Improvement will not come by means of paternal forms of government, for the reason that the socialistic principle is fatal to individual enterprise, and antagonistic to all the influences which can inspire the many to work their way higher. Neither can it be brought about by the promulgation of sentimental doctrines which teach the laborer that he is a poor, weak member of society, who needs guardianship. Everything of this kind increases dependence and discourages personal excellence and ambition.

Rather should we look for improvement wherever the interests of the two elements can be blended and unified; and production be increased, by subduing prejudice and cultivating harmony. Promulgate the fact that the interest of one is the interest of both.

Co-operation has been suggested as a solution of labor troubles. This has merit but has not been uniformly successful. The requisite brain force to organize and conduct business enterprises successfully has often been wanting among working co-operators. But if they can secure a management which combines executive ability and honesty, they may get the advantages of proprietorship. Failing in this, co-operative experiments will not succeed.

A system of profit sharing, by means of a more or less intimate industrial partnership, already considered in a previous chapter, is, however, more promising, and the principle is capable of wide and general application in one form or another. We believe that the escape from present difficulties can only be successfully sought in this direction, for nothing else will weld the two interests that are so popularly supposed to be diverse. The adoption of this plan will require capitalists and employers to take the initiative, which they can well afford to do in view of the prevailing discontent and antagonistic feeling in the ranks of labor. Whether or not there is good foundation for this feeling, it exists, and therefore some movement must be made, or all interests will suffer. These strained relations result in unwilling, imperfect, and lessened production, causing a loss to all interests. Their natural effect upon the laborer is seen in his rendering the least possible service that is compatible with full wages. His heart is not in his work. Give him even a small stimulus besides mere wages, and then note if there be not improvement. Offer to those who are faithful and industrious a bonus at the end of the year. See if a division in this way of five, ten, or fifteen per cent of the profits will not prove mutually beneficial. The employer should be frank and sympathetic with his employees and thus gain and merit their confidence and respect. In the case of railroad employees and other kinds of employment where it is not practicable to divide a percentage of profits, a system of rewards for faithful and continuous service promises good results. The *mutual* interest in the amount and quality of production is the important feature, and this may be attained in a variety of ways, of which the above are but suggestions. In this way, the employer will have interested friends in his service, instead of suspicious opponents working under a temporary truce. This would give strength and cohesion to all legitimate business enterprises. Having harmony for a basis they would be pyramidal in stability. The liability of outside dictation, interference, or strikes, under such conditions, would not be worth mentioning. We earnestly advise employers to try experiments in this direction. It may at first appear that the plans proposed are not strictly in accord with Natural Law; but upon investigation we find that the union between self-interest and self-exertion — not necessarily selfish in a low sense — is a principle inwrought in human nature. In social economics, the laws of mind and of finance must be considered in their connection. They overlap and mingle, and exercise a modifying influence each on the other.

To working-men we suggest that even if you have not the promise of a special dividend or bonus, your true interest is with your employer, and not with outsiders. Your hopes of promotion rest with him. As a rule, it will be for his interest to advance you as your merits and services warrant. A half-hearted service has an injurious moral effect on yourself. If you really *belong* higher than you now are, an opportunity, in accordance with Natural Law, will soon be afforded to rise.

Finally, as in every other department, the inharmonies between buyers and sellers of personal service can only be overcome by a more intelligent recognition of the laws and principles underlying social economics. Openness to *truth* must take the place of hardness and prejudice on both sides. Employers must win the respect — yes, and even the affection — of their help, by fraternal interest and forbearance. Pride on both sides is the great source of friction. By immutable law, good-will always tends to awaken a responsive vibration.

The ideal brotherhood of humanity, which many vainly

hope to galvanize into existence by legislative or artificial expedients, *can* be hastened by the cultivation of the higher principles of the human soul. Whatever growth starts from within is natural, and it will finally penetrate society to its outermost limits.

WEALTH AND ITS UNEQUAL DISTRIBUTION.

> "Order is Heaven's first law; and this confessed,
> Some are, and must be greater than the rest;
> More rich, more wise; but who infers from hence,
> That such are happier, shocks all common sense."
>
> POPE.

> "He heapeth up riches, and knoweth not who shall gather them."
>
> Ps. xxxix. 6.

> "High stations, tumults, but not bliss, create;
> None think the great unhappy but the great."
>
> YOUNG.

> "Man wants but little here below,
> Nor wants that little long."
>
> GOLDSMITH.

> "A man he was to all the country dear,
> And passing rich with forty pounds a year."
>
> IBID.

XV.

WEALTH AND ITS UNEQUAL DISTRIBUTION.

The colossal fortunes that were accumulated during and since the great civil war attract wide attention, and the conclusion is reached that natural economic laws must be faulty, otherwise such marked inequality would not exist. Our decided preference is for a more idealistic condition of society in which, if there were not uniformity, there might, at least, be much less sharply defined extremes.

While, however, with much truth, the present is regarded as an era of great and selfish Mammon worship, a more careful and comparative investigation shows that the tide of human altruism among the possessors of great wealth is rapidly increasing in volume. Especially during the last decade, the amount of private wealth which has been freely devoted to public uses, in the shape of school, college, and university endowments, libraries, hospitals, art museums, scientific equipment, manual training institutes, and college settlements, to say nothing of ordinary charities, has been vast and constantly increasing. The time seems not far distant when the possessor of great wealth who does not recognize his moral obligation to society, and the *privilege* of some kind of ministry, will feel isolated and uncomfortable, if not really ashamed of himself. As the spirit of voluntary benevolence receives the grateful recognition of society, a laudable emulation will doubtless increase it yet more rapidly in future. But to grow it must be voluntary and spontaneous. Warmth, moisture, and a hospitable soil, will turn an acorn into an oak; but the growth is from within, and any forcing from without would be fatal.

Any general movement towards a coercive socialism or governmental confiscation would chill and paralyze the spirit of benevolence, and at the same time stir into action the baser and more selfish elements of human nature. If the rights of legitimate private ownership, which have existed through the entire historic period, indorsed by the highest ethical teaching, are invaded, whether by revolution or through the forms of law, it would indicate a great moral collapse. All beneficence must be voluntary, and would cease to exist with the disappearance of individual ownership.

If the complicated problem of how to bestow large charities without real danger to character, through the growth of dependence and pauperization, could be thoroughly solved, there is little doubt but that benefactions would be speedily multiplied. Every student of social science, as well as every would-be benefactor, appreciates the difficulties which surround this question.

Can fortunes be limited? This is a subject which has elicited much discussion. If it were possible for the government to fix some limit, or a graduated scale of taxation which would amount to a limit, without any violation of personal rights or moral law, could it be made practically operative as human nature is constituted? Let us imagine an attempt to establish a legislative maximum under a possible socialistic economy. Suppose it be placed at fifty millions. There are perhaps a score, more or less, of private fortunes in the United States that would be affected. But others would suggest twenty, ten, five millions, or perhaps one million, as the outside boundary for private ownership. Again, a large number would vote for a hundred thousand, or twenty thousand, and still more for five, and so on down to one thousand, or less, as a final limit. As the improvident and unthrifty are usually in the majority, the proposed standard would vibrate downward, and endless controversy would prevent any final settlement. Let it once be established that a majority, through representative legislation, could vote money from an individual without rendering an equivalent, and where would be the end? There could be none. Any legislative majority, however great, can never really change a natural or moral law by a hair's breadth.

It is not great fortunes, *per se*, that need excite apprehension, but rather the means through which they are accumulated. The great necessity of the times is a revival of thorough honesty, and the sure punishment of its violation. Public sentiment must not applaud "sharp financiering" as "brilliant," but denounce it as socially disgraceful, and punish it as a criminal offence.

By sentimental comparison there is a general feeling of relative poverty on account of existing great private fortunes. Men measure themselves among themselves. But no one is absolutely poorer, but rather richer, on account of existing wealth, even though it be controlled by private ownership. Every social unit in the body-politic is, at least indirectly, better off for general accumulation. It is the human stock in trade, and its lines of relationship extend indefinitely in all directions.

It is a very common but inaccurate saying, that "the rich are growing richer and the poor poorer." A superficial view may give such an impression, but any thorough research shows that the latter part of the assumption is untrue by actual statistics.

There have been changes in general economic conditions within the last thirty-five years, which have incidentally rendered colossal accumulations increasingly easy of attainment. The opportunities afforded by the era of inflation which accompanied the civil war were unprecedented. If the Union had been disrupted, and the currency and obligations of the government not finally brought back to a specie basis, the results would have been far different. Had inflation continued indefinitely, general bankruptcy would have ensued and values largely vanished. Such was actually the case with our neighbors of the South. This shows the application of laws of great significance. Any era of temporary inflation furnishes great opportunities for the bold and sagacious, and for those *already rich*, to add greatly to their possessions. Conditions and tendencies are foreseen and taken advantage of. Property rises in value during periods of inflation; and if succeeding contraction is foreseen, so that it is marketed before the inevitable shrinkage, there is an abnormal profit.

Normal conditions, without either inflation or contraction, are greatly to the advantage of the wage-earner, and all others who have little accumulated wealth. The underlying laws which govern these alternations are so plain, that it seems unaccountable that many would-be "reformers," "populists," and "friends of labor" persistently advocate inflation, and "cheap money," even to the extent of parting company with the great majority of the commercial world. A feverish inflation is the greatest calamity that can happen to the laborer and to all of moderate means. The resulting increment goes to those who already have much *on hand*, for values of existing products are expanded. It is true that if the same property is held through the succeeding contraction, the shrinkage balances the inflation; but the shrewd and far-seeing financier watches the economic horizon, and generally avoids the decline. It is the poor and unintelligent who are the victims of such fluctuations. Business vibrations, even if much less intense than those of the civil war, give the bold and wealthy operator great advantages. They directly kindle unwholesome speculation, and discourage, not only honest industry, but legitimate commerce. An era of inflation or cheap money — so greatly desired at the present time by mistaken enthusiasts — while making some unscrupulous rich persons *richer*, would be a real misfortune to labor, industry, agriculture, and all legitimate business interests. It does, indeed, lend a transient glamour to superficial conditions, but its permanent effects are disappointing and disastrous. It is like the reaction which follows alcoholic intoxication. We ask those who are urging unlimited silver coinage or fiat money — whose motives in many cases we respect — to seriously consider these immutable principles and tendencies.

In periods of inflation wages rise more slowly than the necessities needed by the workman, and, as a rule, salaries for personal service advance only after considerable delay.

Prominent among other more permanent and legitimate causes than inflation, which have made the recent accumulation of great fortunes possible, has been the remarkable expansion of our railroad system. This in rapidity and extent has been unique and entirely unprecedented in the world's history. Within the last two or three decades a territory larger than the aggregate area of all the States east of the Mississippi River has been permeated and developed by the construction of these great public highways. The wealth that has been created by this means ranges in the thousands of millions. To illustrate, take an individual case, and trace the special opportunities afforded for the accumulation of wealth by this great movement. A man with great ability to organize and execute, and with wise forecast, possessed of experience and capital, grasps the boundless possibilities of a sparsely settled and unproductive territory. He foresees that all that is necessary to transform its worthless acres into fruitful farms, and dot them with flourishing towns and villages, is cheap transportation. He projects vast schemes of railroad building and executes them, not as a philanthropist, but as a sagacious business man. He has faith in natural principles, which show him that the result of his venture will be a domain occupied by thousands of thrifty settlers, who will furnish his road with business. As a result of his energy and persistence, and in strict accord with Natural Law, his individual fortune is, perhaps, increased by millions, and he has, if honest, earned his reward. Through his instrumentality there has been added to the capital of the nation, not only the railroad, but many times its value in other products and improvements. Land, before worthless, becomes valuable and productive. Instead of a scanty growth of sage brush, boundless fields of golden grain await the advent of the reaping machine. Where an occasional herd of buffalo was almost the only sign of animal life, numberless droves of cattle and sheep are now seen fattening for shipment, to supply the never-ceasing food demand of the world. In place of vast solitudes broken only by the passing of an emigrant train or an Indian hunter, thousands of brawny farmers and laborers find employment and sustenance. Such a great result is the product of the mental force, possibly, of one man. He has furnished occupation for thousands of workmen who would otherwise be left to overstock the labor market. By the amount of his production he has fairly earned his fortune. While his own wealth has been enhanced, he has caused indirectly a production many times greater. The transaction was only a sale of mental force at such a price as the world was willing to pay. The case supposed is only illustrative, but it is typical of many occurring in real life. In the accomplishment of such results, truth is indeed "stranger than fiction."

Other important means by which the opportunities for making great fortunes have been multiplied are found in the utilization of steam and electricity, and by the great number of inventions. These have changed business methods, and increased in almost geometric progression the practical power and possible achievement of a single individual. Great personal ability, when supplemented by such forces, becomes almost irresistible.

The era just passed has been a transition period. The remarkable change in business conditions and methods has been so rapid, that comparatively few had the foresight and courage to grasp promptly the golden opportunities as they were presented. They were never so numerous and prolific in any past period, and they furnished the special conditions by means of which, perhaps, nine-tenths of the great fortunes have been gathered. Not only the building, but the operating, consolidating, systematizing, and, to some extent, the buying and selling of these great highways have contributed to the result. The flow of general capital into small enterprises of a profitable character is easy and rapid, but in great undertakings it becomes timid and suspicious. This has put a very high premium upon unusual foresight and executive ability.

The two great estates of Astor and Stewart are instances of great accumulation that have taken place outside of these special conditions and opportunities. They represent respectively the departments of real estate and commerce. The Astor estate furnishes, perhaps, the most conspicuous example in this country of what socialistic writers call "unearned increment." But is there practically any such thing? It is a natural law that any unusual opportunities for gain will call out seekers and competitors. If unearned increment is such a prize as we are told, why have not all, or at least more, sagacious men bought land? Simply because they thought there were better investments elsewhere. A careful examination will show that, on an average, a fair interest on the money invested in land, *plus taxes and assessments*, will in the end amount to more than the so-called unearned increment.

There are exceptions to this rule in rapidly growing cities and in newly settled farming regions, but not more than in other kinds of enterprise. This theoretic fallacy may be disposed of by suggesting that, had there been any greater prospect of profit than in other average investments, the shrewd business men of America would long ago have discovered it, and would have invested more in land and less in other objects and occupations. It is probable that even the Astor estate has paid out in taxes and assessments all the natural increase that has taken place, which *is in excess* of a moderate rate of interest on its investments. Land must advance in value very rapidly to outstrip these combined charges. The Stewart estate is an example of what individual brain power, exerted in harmony with Natural Law and by its aid, can accomplish in the domain of commerce and traffic.

The great fortunes that were made in mining, and in mining speculations, belong to an era that culminated several years ago. At present, anything but slow and gradual accumulation in this department is exceptional.

In view of these facts, it seems evident that in most cases the great fortunes were incidental to the unique opportunities presented during the last twenty or thirty years. If these special conditions were temporary in their character, the golden opportunities have largely passed, and fortune making in the future will be slower and more difficult.

In regard to railroad building, nearly all the available territory is now occupied by through or trunk lines, and in future this business will be more confined to the construction of short and comparatively unimportant feeders. The undeveloped territory of our own country is becoming more limited. This will narrow what has been a most prolific field for the rapid enhancement of capital.

It also seems improbable that we can expect any such radical progress in inventions and business methods from the present starting-point as has been made in the past. Better appliances, and a nearer approach towards perfection in the application of steam and electricity, will no doubt be reached; but it is not probable that future improvements will be as radical as those of the last half-century. When a barrel of flour can be carried from Chicago to New York for less than it costs to cart it across either city, it is evident that the process cannot be greatly improved.

Again, as wealth has accumulated, the competition of capital with capital has become more intense. Interest, or the selling rate for the use of capital, has declined nearly sixty per cent. It has gradually fallen from the old standard of six per cent to a point which makes it probable that a two and a half per cent government bond can be floated at par. If the value of wealth be estimated on the basis of its earning power, a million of dollars is now worth less than one-half of what it was thirty years ago. Competition between investors is so great that almost any railroad, which pays five or six per cent dividends on its stock, is in danger of being paralleled.

The general evenness of prices consequent upon telegraphic communication and rapid transportation is another instance of the lessening opportunities for great gains by speculative investments. Important changes in market values are discounted long in advance, and are, therefore, very gradual. Price fluctuations being smaller, successful corners and manipulations become more difficult and infrequent.

The laws of inheritance are also great and constant forces working toward the disintegration and distribution of great estates. In this country, with no law of primogeniture, and where, as a rule, there are several heirs to each estate, its dissolution as a great unit becomes very probable. The Stewart estate, before alluded to, is an example. The longest life is not sufficient for a single individual to absorb more than a minute fraction of the wealth of the community, and, whether more or less, the probabilities are, that at his death it will cease to continue as an organized, accumulative force.

The laws of heredity are also powerful in their wealth-dispersing tendency. While there are exceptions, the sons of very rich men do not commonly inherit the peculiar accumulative ability which characterized their fathers. The dominant and controlling talent is generally greatly modified in the son. Instead of a financier, inclination often leads him to become an artist, a professional man, or still oftener, a gentleman of leisure. In place of the habits acquired by an early economical discipline, are those of an extravagant and luxurious character, incidental to his position. He begins where his father left off; and, in many cases, ends where his father began. Not only the exceptional talent is generally lacking, but the more necessary impelling motive. Most of our millionaires began active life with little or nothing, and were obliged to exercise self-denial and economy which laid the foundation for their future success.

Statistics show that the average life of capital is not equal to the average life of man. It is a prevalent idea that the success which has attended the efforts of the few is due, in a great degree, to chance or luck; but this is a mistaken view. Favorable environment is important, but exceptional financial talent, bringing to its aid the principles of Natural Law, improves and transforms its conditions. The character of environment, therefore, becomes largely a matter of choice, rather than fixed and uncontrollable.

The general average of wealth is higher at present than at any previous time, partially as the result of the special causes already enumerated.

It seems probable that the passion for sudden wealth, which has caused so much unfavorable comment by writers of other nationalities, will diminish as conditions become more stable, and opportunities for rapid gain fewer. The fact that the amount of human happiness has but little connection with the amount of individual wealth will become better appreciated. National life and character have hardly had time to become adjusted to the changed conditions brought about by the rapid expansion before noticed.

All classes, including the poorest, are greatly benefited by the operations of capital. For illustration, the immense Vanderbilt and Gould estates represent most largely individual wealth in railroads and telegraphs. The fact of personal ownership, with its income of four or five per cent on the investment, makes no difference with the great balance that goes directly to labor for service and materials. Every laborer gets as much as if the property belonged to ten thousand stockholders, instead of largely to one. This fact also makes no difference with the productive power of capital in performing the multiform functions of society and commerce. If there be a difference in either direction, the organization and operation are usually more perfect under concentrated control. But, aside from these great public enterprises, there are large investments of a private nature, in the domain of art and luxury. The palace of the rich may excite the envy of the passing laborer, but its value in money has already been disbursed to the mechanics who labored in its construction. Every piece of material has been changed, shaped, and fitted from its condition as raw material by busy workmen, who have thereby had occupation and subsistence.

The great and mischievous fallacy which forms the basis of socialistic literature and sentiment may be summed up in a single sentence; viz., *that all wealth is created by labor, and, therefore, belongs to the laborers who have produced it.* This plausible proposition may also be disposed of as briefly.

The wealth *does* belong to the labor that produced it, but *the larger and more valuable part of this labor was mental.* Socialists ignore brain labor, which by Natural Law is much the more important of the two. Many clergymen, philanthropists, benevolent and sentimental people, who are favorably impressed by some apparently humane and attractive features of socialism, overlook this point. The typical European socialist is intelligent and logical. He is a materialist, and does not believe in mind except as being a manifestation of matter. He therefore ignores mind as a factor in production. Even economists like Smith, Mill, and Ricardo gave little attention to the great part played by brain-force in general production. Their observations were made prior to the present era of great invention, when the influence of mental power was not nearly so predominant. The theory that mental effort is not labor, is too shallow to merit serious consideration. Is not the finished edifice as much the work of the architect as of the mason or carpenter? Does not the student, clergyman, merchant, or inventor labor? On the supposition that wealth is the product of physical labor only, some machines would have very large value when measured by man-power.

Under a government like ours, where all enjoy equal rights, it is a malicious proceeding to foment class feuds and arouse envious passions. It is an abuse of liberty, and its fruit would be tyranny in new and worse forms.

During the time in which capital has decreased fully sixty per cent in earning power, there has been an increase in the productiveness of labor. While wages have increased, their purchasing power has also been enhanced by an average decline in the prices of food, clothing, and other necessities. As a rule, in America we have few idle rich men. An eminent statistical authority has estimated that not more than ten per cent of the population of America have accumulated an amount of property sufficient to enable them to live upon its income without personal exertion in case they so desired.

The aggregate production is much larger, and society richer, by reason of the fact that in accord with Natural Law, labor is intelligently directed and thoroughly organized by the brain power of capital.

While the rights of great wealth, legitimately acquired, cannot be ruthlessly invaded, every rich man owes a great debt to humanity. His unchallenged ownership is a social trusteeship. The passion of accumulation, as an end, is a curse, and its inherent penalty, though often slow, is sure. Great financial ability involves a supreme test of character. Avarice shrivels the human soul. Capital is good for the man who owns it, but if it owns him it becomes tyrannical. Such an one is like a bee submerged in its own honey. To one who occupies a racial rather than a selfish standpoint, great wealth, as a means, is a power and an honor.

THE LAW OF CENTRALIZATION.

"*All roads lead to Rome.*"

"*Westward the course of empire takes its way;
The first four acts already past,
A fifth shall close the drama with the day:
Time's noblest offspring is the last.*"
BISHOP BERKELEY.

"*For he that hath, to him shall be given.*"
MARK iv. 25.

"*Even there, where merchants most do congregate.*'
MERCHANT OF VENICE.

"*There is America, which at this day serves for little more than to amuse you with stories of savage men and uncouth manners, yet shall, before you taste of death, show itself equal to the whole of that commerce which now attracts the envy of the world.*"
BURKE.

XVI.

THE LAW OF CENTRALIZATION.

EVERYTHING has its centre. In every department of human activity there is some localized fountain where forces are gathered, and from which they are radiated. All organic growth, whether vegetal, animal, political, or moral, starts from within and progresses outwards. As the centre of the solar system radiates his light and heat in all directions, so other centres send forth their influences, whether social, economic, political, or ethical. But while constantly giving out, they also powerfully draw, assimilate, and concentrate. Paris, as a centre of fashion, not only sends forth its authoritative modes, but at the same time concentrates more and more of her own special characteristics. Growth once under way, tends to gain in relative momentum. Ten talents are added to ten more easily than one to one. Centralization and specialization are rapidly augmenting forces.

Population is heaping itself up in great cities, and wealth, science, art, and industrial production are responding to the same law. These are the results of "natural selection," and amount to a normal socialism. The great focal points where human interests have converged during the historic period have shifted about. Rome, Florence, Venice, Nuremburg, Vienna, Paris, and London, at different times have been the great fountains of human activity.

But the present movement is quite unlike anything of the past. Invention, rapid transportation, and communication have revolutionized former methods, and the modern metropolis has unique powers and possibilities.

Any careful observer, who has watched the currents of trade in the great commercial centres for some time past, could not have failed to notice a constant tendency towards centralization. It has prevailed not only in American cities and towns, but throughout the commercial world. This proves that it is not in consequence of local or special causes, but the result of influences which operate uniformly in obedience to Natural Law. This conclusion is further confirmed by the fact that it has not been caused by, nor in any way connected with, legislation. We therefore conclude that it is a necessary feature of the present great development of invention and civilization.

The growth of cities has been very marked, and also the expansion of the facilities for production.

There is so much apprehension at present in regard to the possible power of gigantic monopolies, that it is worth while to trace out the working of the natural economic laws which have produced these phenomena, and also their legitimate tendency. The present is an era of monopolies. The fact that a few great firms or corporations in each city, and in each department of business, are able to attract a large and increasing share of the aggregate patronage of the public, is patent to every observer. The Scriptural declaration, that "whosoever hath, to him shall be given," is being literally carried out. For illustration, notice the retail dry-goods trade in any of our great cities. Years ago this business was transacted by a large number of small or moderate-sized establishments, scattered in different neighborhoods. At the present time the greater part is transacted by a few colossal establishments. These great institutions have, in many cases, added building after building, and department after department, until their proportions are of astonishing magnitude. All other branches of trade are more or less under the control of the same natural tendencies. There is also a process of centralization in locality,

THE LAW OF CENTRALIZATION.   189

no less marked. The larger cities, owing to their greater facilities and attractions, and to the ease and rapidity of communication, draw business from the smaller places which was formally under home control. In addition to this, there is a decided grouping of each kind of business in some special locality. There is a dry-goods quarter, a banking quarter, and one for almost every leading department of business. Concentration in locality is added to centralization of capital and enterprise.

The operation of this law in connection with manufacturing is also uniform and strong. New industrial centres are formed in conformity with natural conditions and advantages.

Another manifestation of centripetal law is seen in the growth of cities. At distances somewhat uniform, where railroad systems converge, great commercial centres grow up, each having its quota of tributary territory. Their location and growth are not matters of chance, as many suppose, but entirely in accordance with fixed laws. When one point gets a fair start in advance of its competitors, like a larger magnet, it has increased drawing power. It seems to gain an accelerating momentum, so that any city of given size has four-fold greater growing qualities than one half as large. While the lesser may increase somewhat, it naturally pays tribute to the greater. This is as irresistible as the law of gravitation. The centralizing force that locates special kinds of business in special places is also well defined. The milling industry of Minneapolis, the packing of Chicago, the importing of New York, are examples. Manufacturing, although not so thoroughly confined to single places, has its focal points; as Pittsburg in iron, Lowell, Lawrence, and Fall River in cotton, Paterson in silk, and Trenton in pottery.

At first glance, it looks as if this condition of things, especially in the case of great mercantile concerns, was abnormal and injurious. Admitting that it has aspects of this kind, let us examine carefully its practical working. Imagine a typical American city, with half a million population. Twenty-five years ago its retail dry-goods business was done by a large number of small shops in different localities. Now it is largely monopolized by half a dozen mammoth establishments located almost side by side. What is the effect of this condition of things on the general public, comprising say four hundred and ninety-five thousand out of the half-million people? They show by their action that (unless they are greatly deceived) they find lower prices, greater varieties, and better selections at these great establishments than elsewhere. We are obliged to accept this opinion of the great majority of an intelligent public as conclusive. This disposes of ninety-nine one-hundredths of the entire population so far as dry goods are concerned. The next class, perhaps five thousand persons, who, under former conditions, would be in business for themselves, are now either junior partners, or employed on salaries by these great firms. They lose the net difference, whatever that may be, between the two following positions: on one hand, greater independence and the dignity of proprietorship, but accompanied with uncertainty of success; and, on the other, sure, but moderate success, with more dependence. The fact that but a small proportion of men succeed when in business for themselves, as shown by statistics, will still reduce the net difference so much that, even with this small class, it is doubtful which way the advantage would lie. These two classes comprise everybody except the great firms themselves, whose interests it is not necessary to consider. These great institutions have attained their prominent positions by a regular system of evolution, and are fair illustrations of the "survival of the fittest." Given, a rare combination of capital, executive ability and power to organize, with favorable environment, and we have the conditions of increase almost without limit.

In tracing still further the operation of these laws, let us, for illustration, again briefly notice two great monopolies, which are, perhaps, popularly regarded as the most objectionable of any in this country; viz., the Western Union Telegraph Company and the Standard Oil Company. It is not their private transactions with other companies, but their relations to the general public, that we are now considering. No one is forced to have business relations with them, unless he considers it for his interest. It is in their business relation with the public, as sellers of telegraphic facilities and of oil, that we now look at them; for they have no power otherwise to injure the average American citizen. What is now the Western Union Telegraph Company was formed, as nearly every one is aware, by the consolidation of smaller companies and the absorption of rival but weaker organizations. They were willing and ready to be "absorbed," and were well paid for the operation. The prevalent impression is, that because this business is almost entirely controlled by one great organization, it necessarily becomes a dangerous and powerful monopoly, against which the public has no protection. This prejudice against all great corporations is a characteristic of the present time. There may be more danger in the prejudice, or what may come of it, than in the organizations themselves.

Our safety consists in the fact that the natural laws of supply and demand are sovereign, and that there is no danger of their repeal or suspension. What are the practical facts relating to the telegraph company? It is a seller of telegraphic facilities; and the public, which represents demand, holds the key to the situation. The company can afford to sell its services at a lower rate than half a dozen smaller ones could possibly do. Will it? Yes, in the long run; for self-interest will force it in that direction.

There is what may be called a normal rate for this service; and in case the management make a tariff above this point, demand falls off and profits shrink with as much certainty as they would in case it were put below it. Managers of corporations do not always discover at once how low the equilibrium of such rates lies, and that that point is always the most profitable; but experience is a persistent teacher, and these laws are continually pressing in the right direction, until they vindicate themselves and obstructions are removed. An illustration of the operation of Natural Law in governing demand, is seen in the effect of successive reductions in the rate of postage. Every experiment made by the government in this way has been successful. The increase of business that followed each reduction was so great, that but a very short time elapsed before the net revenue was larger than before. The true normal rate may still be a little below any point yet reached.

The fact that railroad or telegraph corporations have, or have not, "watered their stocks," is popularly supposed to have a great influence on their rates of service. Not in the least. If for any speculative reason the stock of the Western Union Telegraph Company were largely increased or diminished in its nominal amount, the management would find that it would be entirely inexpedient, for that reason, to change its tariffs. Its material facilities would remain as before, and so would the demand for their employment. In other words, the normal point of greatest business and profits would remain the same, regardless of changes in the nominal amount of stock.

We have considered these extreme cases of monopoly, not because we admire or defend them, but only because they furnish another illustration of the supremacy of Natural Law. They may be powerful enough to influence legislation, but they cannot change natural principles.

Their business methods, and dealings with rivals and competing organizations, may have been indefensible, but unvarying natural conditions will make them powerless to harm the humblest American citizen. The man who lets their stock alone, and only buys their product, can never be harmed. They cannot force demand, but only court it.

The most intense and ceaseless competition is that of capital with capital. Often great manufactories run — sometimes for years — not only without profit or interest, but at a small loss of principal, rather than to shut down. To stop is ruinous, because skilled help is scattered, and everything disorganized *except fixed charges*, which go on, and sometimes even increase. The public get the benefit of consequent cheap production. So long as freedom and confidence prevail, idle capital is always waiting to pour into any channel that promises a return of four per cent per annum.

We are led to conclude that the menace to government and to citizens by great business combinations is much overrated. Without regard to legislation, Natural Law hedges them in on every side. While great aggregations of capital, in their operations, are subject to abuses, they are great forces in production, and have an important place in the economic functions of society.

There is every reason to believe that as art, invention, and civilization continue to progress, centralization and concentration will become still more pronounced. . Organization in accord with law is growing more exact and complete. Large cities will grow larger and specialization will be more and more thorough. Each one will do just that which he can do the most perfectly, and thereby make his services to society of the highest value. Wherever cotton, silk, wool, iron, steel, ships, or any other product can be produced or manufactured with the greatest facility and economy, at those points the inherent centripetal and centrifrugal forces will strengthen. Everything will have its distinctive headquarters, and *there* will be concentrated the supremest excellence, adaptation, and economy.

ACTION AND REACTION, OR "BOOMS" AND PANICS.

"*Extremes in Nature equal good produce;*
Extremes in man concur to general use."
POPE.

"*After a storm comes a calm.*"

"*There is in the worst of fortune the best of chances for a happy change.*"
EURIPIDES.

"*There is a tide in the affairs of men,*
Which, taken at the flood, leads on to fortune;
Omitted, all the voyage of their life
Is bound in shallows and in miseries.
On such a full sea are we now afloat;
And we must take the current when it serves,
Or lose our ventures."
JULIUS CÆSAR, Act IV.

"*The time is out of joint.*"
HAMLET, Act I.

"*One extreme follows the other.*"

"*Every white will have its black,*
And every sweet its sour."
SIR CARLINE.

"*In every affair consider what precedes and what follows, and then undertake it.*"
EPICTETUS.

"*Cause and effect, means and ends, seed and fruit, cannot be severed; for the effect already blooms in the cause, the end pre-exists in the means, the fruit in the seed. The changes which break up at short intervals the prosperity of men are advertisements of a nature whose law is growth.*"
EMERSON.

XVII.

ACTION AND REACTION, OR "BOOMS" AND PANICS.

POLARITY is not only an economic but a universal law. Wherever we turn in the broad domain of nature, there are the positive and negative, heat and cold, ebb and flow, general undulation. Excess leads to deficiency, and deficiency to excess. Things that are the most precious, when abused become the most obnoxious. Vibration is continuous both within and outside of human nature. The floods of springtime are followed by the droughts of summer. After activity comes rest; after elevation, depression; after light, darkness.

If we soar above the normal business level at one time, we will certainly fall below it at another; and the higher the flight the greater and more rapid the fall. The most severe panics are generally preceded by intense activity and speculation. The clouds may be slow in gathering, but when they break, like a thunder-storm, they clear the atmosphere.

Every one is aware of alternations of what are popularly known as "good times" and "hard times." But many overlook the fact that they are governed by fixed laws, and regard them as matters of chance, or, at least, as the result of some political or monetary circumstance which has but little to do with their advent.

A panic is a fright or loss of confidence in the stability of existing conditions. There are gradual panics, though the term is more exclusively applied to those which are sudden and unexpected. The antecedents of a depression may be vital and adequate, or possibly exist only in the fancy or supposition of disaster. An alarm of fire in a public gathering may cause a stampede, even if it have no basis of fact. The panic is in the people. But financial alarm is usually the result of causes which give a reasonable ground for apprehension.

Until human nature is evolved to a higher plane there will be flood and ebb tides in the turbulent sea of finance. This would still be true under the most ideal political and monetary system that can be imagined. The feverish desire to get much for little, to gain profit by a short-cut, especially where there are easy facilities for credit and speculation, always leads directly to overtrading which is the sure precursor of shrinkage and disaster. The modern facilities for making large transactions by the deposit of small sums called "margins" furnish a powerful enticement for unhealthful expansion. Artificial attempts to "bear" or "bull" the market lead many into financial deeps until they are submerged. People are never quite prepared for the arrival of a panic, and to many it comes like "a thief in the night."

Overtrading may take place in real estate, stocks, wheat, or tulips. Commodities are only the tools or instruments which are made use of to gratify the desire for rapid and abnormal gain. It is not legitimate industry and commerce, but unhealthful speculation, that brings disaster.

The fundamental and primary condition which results in panic may be expressed in one word, — debt. In itself, debt is not necessarily an evil, but its abuse leads to disaster. An experience of profit leads to larger ventures, and these being successful to still larger, until both individual and collective indebtedness grow to great proportions. When the crisis comes, all want what is due to them, and but few are able to respond. Money becomes scarce and abnormally valuable, and productions unsalable, except at great sacrifice. Business is therefore paralyzed; for all are anxious to sell, and none wish to buy. No human prudence can entirely provide against these convulsions, but a study of their laws and causes may do much to mitigate their severity. A money market always even and in perfect health would imply the prevalence of an almost infallible wisdom, which is nowhere found.

In times of business activity, the fuel is gradually gathered, stick by stick, and added to the pile which is to produce the coming conflagration. When the conditions are ripe, only a spark is necessary to bring general disaster. The proud fabric which has been gradually rising, the stability of which was unquestioned, is dissolved with appalling suddenness.

The tulip mania in Holland, which occurred in 1636-7, is a striking illustration of the possible intensity of baseless speculation and succeeding panic. A single root was sold for thirteen thousand florins. The ownership of a rare bulb was often divided in shares, and many were sold for future delivery by people who did not possess them, and often the article sold was not in existence. The crash came without warning, and was most disastrous and complete. The result was not due in any degree to bank-note expansion, as Holland at that time had only a coin currency.

Laudable undertakings if overdone may issue in panic. The London South Sea Bubble, and some of the railroad panics of America, are examples. They are an evil which no monetary system, however sound, can prevent, and governmental measures are also futile to avert them. With the natural human desire for rapid gain, and convenient facilities for speculation, overtrading is a sure result. It is a peculiar feature that those most actively engaged are less capable of judging of the danger, and the probable time of culmination, than those who look on from the outside. An observer, even in another country, will often discover signs of approaching catastrophe which are overlooked by active participants.

As a rule, important panics are preceded by several years of prosperity which at length reaches a feverish and unhealthy stage. Industry and economy are at a discount, and slow gains unsatisfactory. Production diminishes as speculation increases. Banks expand their circulation and discounts, and individual and public credits are enlarged. Confidence is strong, and profits rapid and large. But at length a day of reckoning comes. Some unexpected weak spot in the financial edifice gives way, and every part comes down, as a row of standing bricks are levelled by the fall of one. Distress, bankruptcy, and liquidation follow; and after a few months, or years, the rubbish is cleared away, and a slow and tedious process of recuperation sets in. Economy again becomes the rule, and extravagance the exception. If the pendulum swings far in the direction of wild speculation, it will go with an equal momentum to the side of depression and stagnation.

Our most notable panics occurred in the years 1822, 1837, 1857, 1873, and 1893. Others of much less intensity, and somewhat different in character, occurred in 1861, 1866, 1884, and 1890. That of 1837 was, perhaps, the most severe of all in its immediate results, and the most lasting in its effects. Ten years passed before values fully recovered and business resumed its normal activity. The principal antecedents were a great expansion of banking and bank credits, and an intense speculation in real estate, especially in New York City. In 1830 there were three hundred and twenty-nine banks in this country, with a capital of $110,000,000. In 1837 they had increased to seven hundred and eighty-eight, with a capital of $290,000,000. Prices of all commodities advanced rapidly, and industry and frugality were at a discount. Many abandoned agricultural pursuits and removed to towns or cities, to speculate in real estate and enjoy their rapidly increasing riches. At length the climax was reached, and the succeeding crisis occurred on May 10, 1837. Careful estimates subsequently made, showed an actual shrinkage of two billions in the value of the assets of the country, and an amount of indebtedness of six hundred millions wiped out by actual bankruptcy. Complete specie resumption by the banks in all the States did not take place until 1843. Thousands who thought themselves wealthy lost all, and had to make a new beginning without a dollar. Labor was a drug, and all property unsalable except at ruinously low prices. Values sunk as much too low as they had before been too high. Recovery was slow and difficult. It required years of toilsome effort to ascend the same hill that had been descended at a single leap.

The panic of 1857 was, perhaps, next in severity, and the preceding conditions were similar. The influx of gold from California, after its discovery in 1848, was added to other speculative elements, and its effect was to intensify the passion for rapid gain. The severe object lesson of twenty years before had been forgotten, and history repeated itself. The prostration was not as severe, and the recovery more rapid than before; but yet the disaster was great, and thousands of fortunes were swept away. The suspension of specie payments by the New York banks, however, lasted only fifty-nine days. Recovery to the normal standard of business and prices was not quite complete in 1860, when the great political events occurred which led to the civil war of 1861. The opening of hostilities produced violent changes and irregularities in our banking system, which precipitated a crisis in the currency. This was quite unlike the panic of 1857, and less severe. The bonds of various Southern States had been largely used in the North as a basis for bank circulation; and as their value rapidly declined, great confusion in our monetary system followed. Financial operations and exchanges were much disturbed, until the exigencies of the war forced the government to issue the greenback currency, which soon took the place of State bank issues. We are dealing with principles, and not with history, and will only briefly notice these monetary changes and their effects. As the war progressed, the redundancy of paper currency increased, and soon caused it to sink below a gold basis. This movement grew still more pronounced when the national banking system was inaugurated, which was another outgrowth of the financial needs of the government. It was devised to aid in making a market for government bonds, which were made a basis for national bank circulation. These issues, added to those of the government, caused a still further depression from a specie basis, until at one time their value was less than half that of gold. A corresponding inflation in all prices occurred, as rapidly as an adjustment could take place, and speculation was the natural accompaniment. As the volume of currency increased, its purchasing power diminished. Supply and demand must come to an equilibrium. There was, however, but little change in prices when measured by the gold standard, the apparent increase in values being in reality fictitious and artificial. Those who were sagacious enough to keep their assets largely in commodities during the expansion profited, in case they turned them into money before the contraction. Thus, we meet the law of supply and demand at every turn, always uniform and supreme. The legislation of the "Medes and Persians" could not excel this principle in unchangeableness. The quantity of circulating medium in any country has a direct relation to the price of its commodities.

The circumstances preceding the panic of 1873 were somewhat different from those before noticed. Its most prominent cause was an abnormal amount of railroad building. This is a laudable business, but it is quite possible to overdo it. There was also an unusual amount of real-estate speculation, and consequent inflation of prices. Whatever single feature may be the direct cause of any panic, its effects spread to other enterprises, even if entirely different in character. As a consequence, other values suffered nearly as much as those of railroad stocks.

By means of debt and inflation, current values of fixed forms of property become too great in proportion to the existing volume of money. The disparity increases until panic comes, which consists of an excited bidding for money by those who must dispose of surplus property. In their competition for money they offer an increasing quantity of commodities for it, which is called a fall in prices. A given sum in this way becomes more valuable, as measured by other property, in accordance with supply and demand.

Alternations of prosperity and adversity will come at intervals, even if the currency, tariff, and all other conditions were the most perfect that could be devised. No matter how liberal the amount of the monetary medium may be, when like a flood-tide the fever of speculative enthusiasm sweeps beyond its normal limit, general apprehension comes, and the inevitable ebb follows. Panics *seem* to come from a lack of money, because it is hard to get — relatively dear — during such periods. The *real* difficulty, however, is the lack of *confidence*. Even with an unprecedented volume of existing currency, when confidence is destroyed money is hoarded so that the supply *appears* to be utterly inadequate. At such times there is a general indisposition to invest and even to pay obligations in the usual way. People seem seized with the impression that if they pay out what money they have they will never get any more. The difficulty is located in human fallibility and not in external conditions.

Disaster generally comes more from anticipated trouble than from that which is actually present. Future conditions are discounted in Wall Street as freely as promissory notes. Sentiment is often as powerful as fact in the regulation of values. Is this due to chance rather than Natural Law? Not at all, for sentiment is a natural law in the human constitution. It is a complex blending of tendencies both within and outside of man that constitutes political economy.

Rarely are panics such unmitigated calamities, or booms such blessings, as they are painted by the human fancy. They have wrapped up in them self-regulative forces which in due time make their power visible.

When everybody is buying it is time to sell, for such a condition cannot continue. The reverse is equally true. When under the influence of a common impulse to sell, values seem to have no bottom, the wise investor will profit by the acquisition of sound properties.

The panic of 1890 was only an echo, though a startling one, of the collapse of the great house of Baring Brothers in London. While not accompanied by any distrust of the currency, it caused an important shrinkage in values and a general stagnation which was slow to mend. Financial disturbances in Australia and South America also cast their shadows over the United States and caused a considerable withdrawal of English investments, all of which tended to retard recovery. An unhealthful over-capitalization and speculation in "industrials" also prevailed between the panics of 1890 and 1893. With that partial exception, the great panic of 1893 was not preceded by the unwholesome inflation in values which usually forms the antecedent of violent disturbances.

The panic of 1893 was distinctly a currency panic. At first glance it would seem anomalous that such a disturbance should come at a time when the volume of currency was unprecedentedly large and constantly increasing. But the inception of the disaster had to do with its quality rather than its quantity. The coinage of 420,000,000 of silver dollars of the sixteen to one standard, during the period from 1878 to 1893, resulted in a depreciation of their bullion value of $175,000,000, or about forty per cent. To this already overweighted currency there was added the monthly coinage of depreciated metal provided for by the "Sherman law," which was causing a steady and persistent inflation. During the early part of 1893 an extensive outflow of gold on foreign balances increased the general apprehension. The redemptive gold reserve was depleted; and serious doubts prevailed, both in Europe and America, as to the ability of the government to maintain the parity of gold and silver. A great drop to a silver basis seemed to be impending. From the nominal ratio of sixteen to one, the bullion value had changed to twenty-eight to one. This state of affairs led to the return from Europe of large amounts of American stocks and bonds, to be realized upon before the apprehended change should take place. The strained situation also led to a general hoarding of gold, which caused a sudden contraction and further loss of confidence. Runs on banking institutions began, and fears of general disaster culminated in such a contraction that currency in small denominations commanded a premium of three to four per cent. There was, in reality, plenty of currency but even more distrust. As money and coinage are more fully considered in a special chapter, these points are only briefly touched upon in this connection as directly bearing upon the notable panic of 1893.

The repeal of the silver purchasing clause of the "Sherman law," Nov. 1, 1893, by the Congress which was convened for that special purpose, restored confidence and assured the commercial world that the existing volume of silver coin would be maintained on a parity with gold. This could only be done by a free exchange of the more precious metal for silver coin whenever demanded. Interchangeability is absolutely necessary to the continuance of equality.

The panic of 1893 was unique in its inception, characteristics, and outcome. It was also peculiar in the fact that legislation was required as a remedial measure. Previous faulty legislation had produced the disturbance, and its repeal was therefore required to allay it.

The economic relations between the leading commercial nations are now so intimate and responsive that any financial disturbance in one sends its corresponding vibrations through all.

It is probable, however, that the panics of the future will be less severe than those of the past. Present business methods and conditions make it improbable that such convulsions as those of 1837 and 1857 will again occur. Rapid communication tends powerfully toward world-wide evenness of prices, and promotes the gradual discounting of what would otherwise be violent fluctuations. There is also a growing sentiment against excessive individual indebtedness, and business is more generally conducted on a cash basis. International commerce also conduces to steadiness of prices, and any abnormal prosperity or depression in one country receives a corrective influence from others. There is a better understanding of Natural Law, and a more general appreciation of the certainty of the penalties for its violation. When all are familiar with unerring natural principles, and have confidence in their continuous operation, they will become less susceptible to such impulses as issue in a financial crisis. When exciting and disquieting rumors prevail, even the strongest will sometimes lose their equanimity. Anything like a stampede in the financial world is disastrous. Reassuring influences are very necessary at such times. Often a firm and united stand taken by the banks, with mutual assistance when necessary, accompanied by a temporary increase of circulation, or an issue of clearing-house certificates, will alleviate the worst features of an economic convulsion. A subsequent steady and slow contraction on the part of the banks, after the excitement subsides, will generally take place, to conform to the changed business conditions. The greatly increased general foresight in determining the future tendency of market prices will do much to prevent any repetition of severe panics, for dangers foreseen can be largely avoided. Steady and even markets do not present good opportunities for speculation and rapid accumulations by the unscrupulous, but are favorable for labor and all legitimate business and industry.

MONEY AND COINAGE.

"For what is worth in anything
But so much money as 'twill bring?"
BUTLER.

"Money alone sets all the world in motion."
PUBLIUS.

"The love of money is the root of all evil."
1 TIMOTHY vi. 10.

"Remember that time is money."
FRANKLIN.

"Silver and gold are not the only coin; virtue too passes current all over the world."
EURIPIDES.

"It is a condition which confronts us — not a theory."
GROVER CLEVELAND.

"Gold! Gold! Gold! Gold!
Bright and yellow, hard and cold."
HOOD.

XVIII.

MONEY AND COINAGE.

THERE is perhaps no other department of economics where so many and diverse theories prevail, as in that which pertains to the monetary circulating medium. The endless profusion of papers, opinions, and Congressional speeches that have been put forth by impracticable theorists during the recent past, tends to muddle and complicate principles which in themselves are natural and simple. Prejudice, partisanship, and supposed sectional interests color personal opinion, and the result is seen in distorted and fragmentary views of natural unchanging principles. An imagined diversity of interest of sections and parties is fatal to a search after truth, for that is *always* unitary. It would be as absurd to suppose that the same cause could produce health in the right side of the human organism and disease in the left, as to believe that any monetary policy could at the same time have a prosperous and adverse effect in different parts of one nation. *Diversity of interest is only on the surface.*

It is obviously beyond the scope of this work to consider technically, in a single chapter, a subject that alone would fill a larger volume if exhaustively treated. Historical and statistical aspects are of interest to the student of economic science, but the average reader will find more profit in a concise study of inherent and fundamental principles. Let us make a judicial search for the uncolored truth.

What has Natural Law to do with money and coinage? Does it give any indication as to the kind and quality of a normal or ideal currency? Does it shed any light upon bi-metallism or monometallism, gold, silver, nickle, or copper? Yes, it touches these questions on every side.

The present circulating medium is the result of a long evolutionary process. In Colonial times tobacco was a legal tender currency in Maryland and Virginia. In other colonies, coon skins, beaver skins, and musket balls were extensively used as money. As late as 1866 hand-made nails passed current in some of the secluded villages of France. From 1785 to 1787 a limited coinage of cents and half cents was made by private parties in several States under the authority of legislative Acts. In 1786 Massachusetts passed an Act to establish a State mint for the coinage of cents and half cents. The United States mint was not established until 1792.

It is natural that there should be some current representative of value. The most perfect system of barter is unsuited to any condition above the plane of barbarism. Money represents stored-up labor, otherwise it would be valueless.

Conceding the necessity of a currency, what should be its qualities? The most important requisite is stability of value. So far as is possible, it must be rendered independent of fluctuation in value and volume. To have a currency constantly liable to grow cheaper or dearer, either through the changing temper of legislation, or variations in the bullion value of metals, is prejudicial to legitimate industry and commerce.

Steadiness being the great desideratum of an ideal currency, what are its factors? Objectively they are two, and they are often so blended that a distinct line can hardly be drawn between them. The first is the intrinsic, or natural element — which consists of the stored-up labor embodied in bullion — and the second is the fiat of legislation.

An illustration of the natural element is found in the early history of California. Immediately subsequent to the gold discoveries of 1849 the currency consisted of that metal, first in the form of dust, and afterwards of private assay put in the general form of coins. The stamp of well-known and responsible firms upon a piece of gold of regulation weight and fineness made it locally current as a monetary medium. Being of equal bullion value with governmental coinage, it circulated for some time side by side with it, until a full supply of the latter finally displaced it. It was not a counterfeit, for it did not resemble the national issues, but having a full natural value it did not require any of the artificial or fiat element.

The governmental stamp upon a disk of metal is merely a certificate of its weight and fineness. In itself it adds no value except as it confers the privilege of an interchange with something else of higher value. For example, the present silver dollars of the United States, even though bearing the official device, would soon sink to their *natural* value were not the public honor pledged to keep them at a parity with gold through interchangeability. So long as an inferior thing can be freely exchanged for a superior, it will be its equal through the aid of an artificial element, *which consists of that privilege.*

Some would-be political economists place great stress upon the artificial element in currency, and consider the governmental fiat fully adequate to create value. Not so. A purely fiat currency might have *some* value by being made receivable for national dues; but lacking a redemptive basis, it could never be otherwise than artificial and uncertain. Unless kept in very restricted volume, it would invariably depreciate from the real monetary standard. It would lack any solid commercial foundation. Value must be earned, and cannot be created even by a great nation. It must possess stored-up labor. It *is* value only from the fact that it *costs* something.

The precious metals, on the basis of natural cost and general demand, have the same value throughout the commercial world, but within the limits of different governments the artificial element of legislation, or legal-tender value, is blended in varying degree. But this is all eliminated, and cuts no figure, in international commerce and exchanges.

It may be admitted that the same volume of currency changes in potency in special seasons of prosperity or adversity; but this is not the fault of the money, but of the uncertain moods of humanity.

Besides the two objective elements of value — the natural and artificial — already noted, there is also a very important subjective factor known as confidence. It is an invisible currency in itself, for it pieces out that which is in sight. It is never lacking in its connection with natural value, but it fluctuates greatly in its combination with that which is legislative or artificial. Its scarcity often proves a greater calamity than the actual deficiency of its material counterpart.

In times of panic and depression, superficial observers locate all the ills in the lack of a sufficient circulating medium, while often the *only* want is confidence. People are doubtful about the stability of the artificial conditions. When there is an apprehension that the existing grades of cheaper money, whether silver or paper, may break away from an equality with the standard, in spite of the efforts of the banks or government, there is a *currency* panic. Confidence is the barometer that is always testing the future of the artificial element in money.

In more primitive times, owing to permanent distrust, transactions were at once closed with real rather than representative money. But with the growth of the modern commercial spirit and a higher civilization, there has been a great and continuous increase of confidence. This has formed a basis for *credit*, through the use of which both domestic and international commerce has enormously increased. It has also furnished a field for the universal use of bills of exchange, drafts, checks, and other representatives of real money. So long as there is no doubt in regard to ultimate redemptability, almost an unlimited amount of business can be transacted with monetary representatives. But with the prevalence of any unsound financial theories, of present or prospective faulty legislation, or of any sort of departure from conservative solidity as to *basis*, apprehension begins, and real money is demanded because its representatives are distrusted. This leads to panic and subsequent stagnation, as illustrated in 1893. With an undoubted currency and a prevailing sense of moral trusteeship in corporate management, "good times" would be perpetual. Confidence forms the key-stone in the arch of prosperity.

The more perfect the banking system and the sounder the financial policy of any country, the greater is the amount of business that can be transacted on any given reserve of real money. The balances settled in a single week at the New York Clearing House are often more than the entire circulating medium of the nation.

The world's international currency — which represents gold — consists of bills of exchange; that of domestic wholesale trade and banking transactions, of drafts and checks, and even of the domestic retail trade only a small part is transacted by the means of real, or basal money.

The evolution of the banking system of the United States during the last thirty-five years has been remarkable. With all the existing popular prejudice against national banks, they are a vast improvement over any system of the past. It is true that the amount of governmental bonds which forms their basis is diminishing, so that some important modifications may be necessary in the near future, nevertheless they have served a great purpose.

Previous to the civil war there were as many kinds of currency as there were States. The bank issues of many sections were uncurrent outside of their own sectional territory, and only salable to brokers at a discount. In many cases their security was precarious, the chances for redemption in any considerable amount doubtful, and their ultimate valuation problematical. Counterfeits were numerous and difficult of detection, often requiring the services of an expert. The direct losses to holders of this heterogeneous currency amounted to millions yearly, to say nothing of the general business derangement which resulted.

The present uniform value of the National bank issues throughout the length and breadth of the country, the positive security to bill holders, even in the case of occasional bank failures, and the immunity from counterfeits, altogether form a striking contrast with former conditions. Whether or not this system will gradually be superseded by national issues and a sub-treasury plan, is not a vital question in its relation to Natural Law. Circumstances alone will determine the relative preponderance of the two systems. Both accord with sound principles so long as an adequate reserve of standard money is pledged to their redemption. Under either system sharp fluctuations in the volume of the circulating representatives of money must be avoided.

National banks are considered as a "monopoly" by some; but so long as they are organized under general laws and not special "Acts," it is difficult to see the propriety of such a characterization. The fact that there is some voluntary retirement among them, and that their average dividends do not now excel other well-managed enterprises, does not comport with popular impressions regarding their exceptional advantages for profit. Iconoclasm is sometimes useful, but a disposition to pull down tested institutions with the idea of replacing them with financial vagaries is unwise.

Turning from money in its broad sense to basal money — or coin — we find that through the entire historic period gold and silver have been the chosen embodiments of value among nations of any considerable degree of civilization. Natural selection has taken them from among the metals for such a function because the labor involved in their production renders them both scarce and precious. This confers large value in compact form — more especially in the case of gold — and it is further enhanced by their beauty and comparative incorruptibility. Their practical utility, however, in hand-to-hand transactions is still limited by weight and bulk as compared with their available representatives. The important abrasion of gold is also a serious obstacle to its constant use. The coined metal, therefore, mainly rests in bank vaults or depositories, while its more portable agents — printed notes — perform its function. In modern commerce, however, its office is still more largely delegated to checks, drafts, and bills of exchange. These, though temporary, are representatives of current money, as bank notes are representatives of real money, or standard coin. The active office or duty of coin is therefore almost entirely delegated.

There are two different ways of measuring the value of money. One is by the varying amount of products that it will command, and the other by its degree of usefulness as shown in the current rate of interest. The fallacious theory that, owing to the lessening use of silver, the existing currency of the world is growing dearer, is doubly refuted. First, by the increasing amount of *average* products that it will command, and, second, by the very important decline in the rate of interest. The latter in any free market represents the general consensus of opinion as to its desirableness and possible utility. *Interest could not decline if money were not relatively more plentiful.* Bids for the use of money must be as exact and conclusive as rentals offered for houses or farms. Supply and demand in each and all cases is the arbiter. As indicated by the dual test of products and interest, the circulating medium, of whatever composed, has been steadily growing cheaper. The same is true as measured by wages, taking the average of all grades. These facts are noted in refutation of some popular prevailing sectional theories.

Assuming that gold and silver compose the normal and universally accepted coin-currency, what is the teaching of Natural Law concerning them? First, that in reality there cannot be two *standards*, any more than there can be two yardsticks. Bimetallism is often regarded as implying two *standards*, but it only signifies two *metals*. Bimetallism is a rational and practical accomplishment, but two different sized measures are mathematically impossible. As well different-sized bushels, or lighter and heavier pound weights. To keep the distinction between two metals and two standards clearly in mind, will aid in the study of this much-befogged problem.

Natural Law and evolution would indicate that with the immense modern accumulation of wealth and the greatly augmented volume of commerce, there would be a corresponding tendency toward a more valuable and concentrated monetary basis. Among barbarous tribes the currency is composed of beads and shells; and, as ascending steps in civilization are taken, iron, copper, nickel, silver, and gold come respectively into relative use in an advancing order of value. Evolution is a universal principle. Things that are cumbersome and inefficient are continually being displaced by those of greater perfection. There is really nothing more strange or revolutionary in the basal substitution of gold for silver, than in that of electricity for horse-power, or railroad service for that of the stage-coach. Where wages are not more than ten or fifteen cents per day, as in China, small copper coins of trifling value can still be utilized; but with the advance of values and civilization, improved tools and instruments are a natural accompaniment.

The ratio of valuation between gold and silver bullion

MONEY AND COINAGE.

has vibrated somewhat during the whole historic period, but never before has the divergence been so wide as at the present time. As in every other department, the relations are entirely governed by supply and demand.

Herodotus estimates the ratio of gold to silver as 1 to 13, Plato as 1 to 12, Menander as 1 to 10, and in Cæsar's time it was 1 to 9. For some time previous to the discovery of the rich silver mines of Potosi, in the sixteenth century, the ratio was about 1 to 11¼, and in the early part of the seventeenth century, 1 to 12¼. In the present century, previous to the great gold discoveries of California, it was between 15⅝ and 16; and afterwards, in spite of the enormous increase in gold, it only fell back to 15.41.

The following table, estimated from reports of the Mint Bureau, will show the remarkable change in the ratio of gold and silver that has taken place since 1873:—

CALENDAR YEAR.	VALUE OF FINE OUNCE AT AVERAGE QUOTATION.	BULLION VALUE OF U. S. SILVER DOLLAR.	GOLD RATIO.
1873	$1.30	$1.004	15.9
1874	1.28	.989	16.2
1875	1.25	.96	16.6
1876	1.16	.89	17.9
1877	1.20	.92	17.2
1878	1.15	.89	17.9
1879	1.12	.869	18.4
1880	1.14	.886	18
1881	1.14	.886	18
1882	1.13	.878	18.2
1883	1.11	.868	18.6
1884	1.11	.868	18.6
1885	1.06	.82	19.4
1886	.99	.769	20.8
1887	.98	.757	21.1
1888	.94	.727	22
1889	.93	.72	22.3
1890	1.05	.809	19.7
1891	.99	.76	20.9
1892	.87	.67	23.7
1893[1]	.81	.625	25.5
1893[2]	.71	.56	28.4

1. Average first 8 months.   2. November 1st.

The remarkable change as shown above is due, not only to the recent discoveries of enormous deposits of silver, but also to the discontinuance of silver coinage by the leading nations of Europe, and recently by the United States. Germany discontinued silver coinage in 1870, at the close of her war with France, and gradually put her enormous stock of that metal upon the markets of the world. This was like taking a heavy weight from one scale of a balance and placing it in the other, thereby doubly changing the relation. The Latin Union also joined in the same movement, and the departure of silver from its nominal and legislative ratio became more and more pronounced. For some time prior to the action of the United States in 1893, Europe was virtually out of the market for silver bullion for coinage purposes. The currency panic of 1893 was caused by the apprehension of the inability of the United States to maintain a parity even under limited coinage provided for under the "Sherman law."

The Old World for three years had been returning our securities, fearing that a drop to a silver basis might be imminent. Thus the crisis came, not because of too little money, but from the fear that an increasing redundancy might at any time result in depreciation. Every one wished to stand from under what was impending. The world-wide trend was yet further indicated by the fact that India, which had always furnished a very extensive market for silver, joined in the general movement for a gold basis in the autumn of 1893.

It has for some time been evident that nothing less than a general international agreement would so increase the demand for silver as to gradually close the great chasm which now exists between its actual and nominal value. But it is even doubtful whether or not a great international combination would be able to permanently restore the former ratio. Should it at once make the effort to artificially raise silver so much above its natural level, the production of that metal would be immensely stimulated. It is quite probable that in a few years the artificial would again have to yield to the natural. Even nations cannot prevent this. Furthermore, as has already been intimated, such a step would seem to be backward from an evolutionary standpoint.

It is doubtless possible to maintain a parity of gold and silver coin in the United States, so long as no further additions of the latter are made, and thus practical bimetallism can be continued. But with any impairment of the public confidence, either as to the ability or intention of the Government to do this, a process of gold-hoarding could not be prevented. A minority in Congress, who with great persistence advocated the "free and unlimited coinage of silver," upon the former, or a slightly increased ratio, call themselves bimetallists, but it is difficult to see the propriety of such a designation. Such a measure would inevitably bring about silver monometallism. This was so clearly apparent, even under the restricted coinage of the "Sherman law," that the panic of 1893 was the result. It is hardly necessary to suggest that the moment that gold commands any premium, either through apprehension or real scarcity, it will cease to be currency and disappear. This would amount to a sudden contraction of the available circulating medium, but that would be but one of the many disasters which would result from a drop to a silver standard.

No metal can really become a standard unless it possesses international acceptability. There are no walls between nations, and the commercial world is virtually a greater unit. Under modern conditions different countries are neighbors, and no one can disregard the action of the others.

The "silver question" has no class, partisan, or sectional

significance. Aside from the few owners of silver mines, the *whole* country would suffer and become financially disorganized by a drop to silver monometallism. This is as true of the poor as of the rich; of labor as of capital; of the agriculturalist as of the manufacturer, and of Colorado as of New York. In cases of general inflation, as during the civil war, wages and salaries are always the last things to rise to the full proportion of material products. A sound and stable financial system is advantageous to *all* sections. Any theoretic diversity of local interests is the result either of demagogism or of ignorance.

Suppose for the sake of argument that it would be a temporary advantage for a "debtor section" — if such can be truly said to exist — to pay its obligations in a cheapened currency, it would finally prove a very mistaken policy. A high credit is vastly more valuable than temporary profit if such were possible. This is particularly true of the newer States where capital is needed for development. The temper of public opinion as represented in the legislation of each State determines its credit and standing in the financial world. It is for the interest of States to keep their credit so high that both they and their inhabitants can make loans at low rates of interest. Capital is attracted to such localities and becomes cheap and plentiful. Every degree of the element of *doubt* adds directly to the rate of interest in an increasing ratio.

Nothing could be more permanently harmful to the debtors of a State than special legislation which is theoretically in their favor.

It is often claimed, and with some plausibility, that within the last twenty years gold has grown abnormally dear at the same time that silver has been cheapening. But even if this were abstractly true, the practical fact remains that the general currency — which has gold for its basis — has cheapened. This was noted in the early part of this chapter as proved by the average advancement in products and wages, and also by the decline in rates of interest. With the great modern utilization of the various representatives of money, a vastly greater business can be transacted upon any given amount of the ultimate standard than in the past. In domestic commerce, coin cuts but a small figure, and in international transactions it is only used for balances.

But the confidence in coin representatives, as to their *ability* for redemption, must be unlimited. Confidence is the great "power house" of the business world. If all distrust of the currency and of labor friction could at once be eliminated, an era of prosperity, natural and solid, would come to remain.

If the present supply of gold is in any degree inadequate for the basal monetary standard, that fact will stimulate its production the world over. The available supply of the native metal is inexhaustible, and demand always brings supply. With scientific mining methods generally adopted, experts assert that the annual product can be speedily doubled. At the same time the cheapening of silver will greatly lessen its production, and this will prevent an indefinite continuance of the very rapid decline of the past decade.

While but one monetary *standard* is possible, so long as the divergence between gold and silver did not amount to more than one or two per cent, a practical, though not a mathematical double standard apparently continued. But that time has past with no prospect of return.

With gold as the basal standard — even with silver so greatly depreciated — a large volume of the latter can be floated and utilized at a parity. But the amount of silver must never become so excessive as to cause any doubt regarding their free interchangeability. The volume of inferior coined metal that can be utilized in any country depends much upon the customs of the people. The extensive use of a silver coin currency among the peasantry of France enables that country to float a large amount per capita, while in the United States it cannot be extensively utilized except through its representatives.

By Natural Law, there is but one way to provide for bimetallism in any country; and that is to make the more precious metal the standard, and then float such an amount of the cheaper metal as can be kept upon an undoubted equality through free interchangeability. If an attempt be made to make the cheaper of two metals the *standard*, the dearer, under all possible circumstances, will disappear.

Currency panics are inevitable if any element of doubt exists as to prompt redemption whenever required. With prevailing unimpaired confidence the only use of the coin standard is for redemptive reserves and foreign balances. It is like a yardstick that is only occasionally brought out to verify the professed length or width of fabrics.

The principles regarding money and coinage that have been outlined are true because they are natural. With the elimination of the artificial elements that have been injected into the subject by sectionalism, partisanship, and one-sided aspects, difficulties vanish, and unity and harmony are seen to be reasonable.

TARIFFS AND PROTECTION.

"*Idleness and pride tax with a heavier hand than kings and parliaments. If we can get rid of the former, we may easily bear the latter.*"

FRANKLIN.

"*He smote the rock of the national resources, and abundant streams of revenue gushed forth.*"

WEBSTER ON HAMILTON.

"*All government — indeed, every human benefit and enjoyment, every virtue and every prudent act — is founded on compromise and barter.*"

BURKE.

XIX.

TARIFFS AND PROTECTION.

A BRIEF study of the relation of tariffs to Natural Law seems proper, but any partisan or dogmatic treatment of the subject would be entirely foreign to the spirit and purpose of this work. Underlying principles can be intelligently traced out only by an unbiased search for truth for its own sake.

The formulation of a customs tariff is a work so complex and many-sided, that it requires both impartial and expert treatment, but even with the best of these the result is imperfect. A tariff is an elastic expression of national *policy*, and is based upon conditions which are constantly changing, therefore it has none of the exactness of Natural Law, though it has relations with it.

As a question of party politics it is warped and twisted by the partisan press, and the average politician finds it difficult to see more than one side. Being purely a practical economic problem, requiring the impartial study of the best financial and ethical talent, it is unfortunate that it has a political mask fastened upon it. Partisan prejudice and expediency directly prevent the very thing most necessary — a calm and judicial study to determine the greatest good for the greatest number. So long as it continues to be a political shibboleth, prejudice rather than unmixed truth will be the determining feature. Opposite aspects of the question are dwelt upon out of proportion, until the existing tariff, whatever it may be, to different observers, is made responsible for all prevailing good and all existing ills. Men of undoubted integrity and patriotism almost become willing that the whole country should suffer, in order that some party policy may be vindicated, political capital secured, and the opportunity afforded to declare, "I told you so." The "outs" attribute all calamities to the tariff of the "ins," and too much or too little tariff is made the "scapegoat" for the sins of the nation. Few think of studying the subject until they put on a pair of partisan glasses. The "outs" feel that the salvation of the country depends upon their becoming "ins." Facts are made to bend to theories until they reach the snapping point. As a matter of scientific research, it would be interesting to determine by what mental process able and honorable men and newspapers on both sides become psychologized by party platforms and prejudices. Some great Republican, Democratic, or other political psychic wave rolls over the country and lifts men off their feet, and they become as children in its embrace.

A customs tariff, whether higher or lower, is a less important factor in a nation's welfare than is generally supposed. Business conditions are elastic, and are not long in adjusting themselves to a revenue system. But a very imperfect tariff which is permanent, *and known* to be so, is preferable to any schedule which is prospectively changeable. One of the great drawbacks to commercial prosperity is the almost continual apprehension of changes in the revenue system.

All tariffs are artificial, and all are obstructions to the free courses of trade and commerce. While this is true, they may be expedient and politic. Aside from the protective element, they are the most natural ways and means for raising a national revenue. They are less cumbersome and more popular than direct taxation, and in varying degree the foreign producer also contributes toward the desired result.

Tariffs may be framed for revenue only, for revenue

with incidental protection, or for revenue and protection. If revenue *only* be desired, the object is most easily accomplished by the imposition of duties upon articles of general consumption that are not of possible domestic production. For example, tea is incapable of home growth, and being an article of almost universal consumption, a very large revenue would be possible by the imposition of a moderate duty. As a revenue producer, a high tariff is often less successful than a lower one, because the former tends to limit consumption.

Revenue with incidental protection would embrace articles of both domestic and foreign production, the prices of which are somewhat enhanced by the duties imposed. The protective element is the largest in the duties upon those things that are capable of unlimited home production, and that also are largely made elsewhere. The present tariff — 1890 and 1893 — was designed both for revenue and protection; but as the revenue was ample when it was adopted, as a rule, articles incapable of home production come in under it free, or at nominal rates.

As at present organized, the two great parties of the country — the Republican and Democratic — represent, respectively, distinct protection with revenue, and revenue with some incidental protection. This can only be taken as an average statement, as indicated by personal exponents and political platforms. In the Republican ranks, opinions are shaded from high to very moderate protection, and among Democrats from moderate protection to free trade.

The brief study here proposed of a few fundamental principles underlying the revenue system is thoroughly unpartisan in intention. We think that each party places undue emphasis upon certain phases of this complicated and ever recurring problem. The importance of the element of American labor values is underestimated by many Democrats; and, on the other hand, some Republicans greatly overestimate the potency of a protective tariff, and forget that it is purely a policy, and lacks the basis of a universal principle. The question of an American tariff is only a question of American expediency.

Could a commission of economic experts be formed with a single aim for justice and the public welfare, occupying an American standpoint, and uninfluenced by political ties and questions of party advantage, they would be able to outline a very perfect revenue system. But with existing partisan prejudice and unsound theory among legislators, and motives of personal, political, and sectional policy, the obstacles to an ideal result are great. Added to the difficulties already enumerated is the general fact that any tariff bill that can possibly receive the support of a majority in the Congress must naturally embody a general compromise.

It may be noted, as a rule, that the adoption of any new schedule is generally fatal to the continuance in power of the party that is responsible for it. Numberless concessions and compromises have to be granted and a great variety of sectional interests placated. It thus becomes a system of shreds and patches, without consistent unity. As no section or interest gets quite all it asks, there is general dissatisfaction. Besides, the "outs" — whoever they may be — pick innumerable flaws; and before time has permitted a thorough test, a reactionary wave results in a vacation for the dominant party. All this produces a restlessness unfavorable to business adjustment. A tariff, once intelligently adopted, should embody a fixed policy, to be depended upon. Under such a plan but few and slight modifications would be necessary during a decade. To make tariff-policy a party "football," is to sacrifice general prosperity to political vagaries. It cannot be denied that both parties share in the responsibility of keeping up this interminable unrest.

The presence of the protective element in a revenue system can only be approved when there is some peculiar local or national condition to be maintained. Without such a reason for its presence, it seems to savor of exclusiveness, if not of unfriendliness. Among the nations of Europe, in their relations to each other, there appears to be little valid reason for a protective accompaniment to revenue systems. Labor values and general conditions being not very unlike, any special measure of protection seems illogical. Protection, in its nature being an artificial intervention, should always have a sound reason for its employment.

Has the United States valid ground for some general system of protection, in connection with its duties for revenue? From a purely cosmopolitan standpoint, no; but from a national point of view we think it has. The one reason why American protection should be moderately imposed against the nations of the Old World, is involved in the question of labor-values.

It is everywhere admitted that the average workman of America enjoys a distinctly higher standard of living than his European brother. As a rule, he is better educated, of higher tastes, accustomed to greater comforts and more privileges, therefore his requirements are enlarged. What will satisfy his foreign competitor will not satisfy him. The fact that he is a sovereign in a land of freedom and political equality, that he is not bound to any fixed class, and has individual aspirations, also enhances his material demands. Through the influence of superior environment he is on a higher plane than the equally skilled foreign producer. Can he be kept there without artificial aid?

The instruments that facilitate production, under the great advantages offered by modern transportation, are becoming very evenly distributed. Any important inventions or labor-saving machines, even if of American origin, are patented and utilized in European countries almost simultaneously with their appearance here. Raw materials, so called, are constantly approximating to general evenness of value. To offset the difference in transportation, the Old World also has some advantages in lower rates of interest, more capital, organization, and specialization. A large number of our workingmen — now American citizens — came from the Old World; but, though some conditions may be more favorable here, it cannot be claimed that a mere change of residence has materially increased their productive ability. In view of all these facts, there seems to be but one way to maintain a distinction, and that is, through reasonable though conservative protection. Water will gain a level unless some kind of a barrier prevents it.

It is not probable, however, that in the absence of all protection, American wages would decline fully to the European level, but they would approximate. A newer country, naturally affording greater opportunity and enterprise, always possesses advantages for the workingman, other things being equal; but as time passes, the gradual tendency is towards equalization.

Suppose that two factories — one in Old and the other in New England — are producing the same kind of goods for the American market. The English proprietor with equal enterprise obtains all the most approved appliances, and his raw material is approximately of the same cost. But suppose the American proprietor pays his help two dollars per day, while the Englishman gets his for one. Can this continue unless the English product is made to contribute a part of the difference? Such a disparity in labor value is constantly appearing in numberless forms. If all bars were taken down, could the American manufacturer successfully compete with his foreign rival unless he be able to obtain labor at about the same price? Having attempted to state the proposition fairly, we leave the solution to the reader's logic.

But protection is neither a full panacea, nor a tool where the edge is all on one side. A bigoted protectionist is as illogical as a free-trade doctrinaire. Protective measures must be employed with great care, otherwise they overreach their usefulness. There is always difficulty in maintaining artificiality. Though the American market — embracing nearly 70,000,000 of the greatest consumers in the world — is of immense proportions, our manufacturers are constantly reaching out to demands that are world-wide. This presents another phase of the question. To aid such enterprising and laudable efforts, the duties upon necessary raw materials must be removed, or at least very delicately adjusted. Principles of possible reciprocity should also be carefully considered. No general rules are possible, for among all artificial conditions each must be separately considered upon its merits. The fact, too, that nations, like individuals, have their prejudices, should not be overlooked. Any tariff which seems partial or specially unfriendly is likely to provoke retaliation. The possibilities of a growing reciprocity are worthy of the study of statesmen and economists.

Does the agricultural producer lose by the slightly enhanced prices he may have to pay for certain products in consequence of their being moderately protected? Not if the domestic market for the fruits of his own toil is broadened and improved, for this may more than offset the difference.

No tariff in itself, however wisely constructed, will make "good times." In fact, unless it be intelligently adjusted, it may prove an unhealthful stimulant. If it be so abnormal as to cause any rapid and undue domestic expansion or over-production, severe reaction will ensue. A wholesome principle that is overdone becomes unwholesome. This law should be borne in mind by partisans on both sides. There are so many silent and unseen compensatory forces at work under any probable system of duties, that dogmatism upon the subject is unprofitable. Radical differences among the most eminent and conscientious observers attest its intricacy and complexity. It is not the creation of Natural Law, and that fact gives great room for theorizing.

The subject of raw material has already been touched upon. But there is considerable difficulty found in an exact definition of that term. Wool is "raw material" to the manufacturer, but the farmer considers it as finished product. If it be made free, the wool-producer feels that he is brought into direct competition with cheap-labor countries, and therefore unfairly treated. Thus endless frictions crop out with any possible tariff. Every special industry, town, section, and State is dissatisfied unless *its* peculiar interests are specially considered. Thus, in the words of a former presidential candidate, the tariff becomes largely "a local issue." A coal producing section desires coal protection, while coal consumers urge its free entry. Trenton would like to see pottery well taken care of, and Patterson, silk, and so on indefinitely.

Everyone is aware that there are two kinds of duties levied on imports, known respectively as specific and *ad valorem*. In many cases the two are combined, thus making the customs duties complicated and cumbersome. Specific duties are much more simple than those based upon valuation. Opportunities for inequality and even fraud are often found through inexact appraisement or under-valuation. Whether higher or lower, the simplification of the American tariff is in the highest degree desirable.

As a matter of history, the protective principle has often been of practical benefit in the incipient stage or early development of special industries. Continued for a few years it has sometimes enabled them to grow from feeble beginnings to enterprises of great importance. In some such cases production has become so perfect that the necessity for its continuance has been outgrown. Consumers have reaped a benefit by being able to get better goods at lower prices than would previously have been possible with free entry.

A few of the general principles involved in the formulation of a tariff may be concisely hinted at as follows. Is the article capable of unlimited home production, so that domestic competition will protect the consumer? If so, and not yet developed, will the temporary disadvantage of the consumer work to his interest in the long run? If an article, like sugar for instance, is only capable of *limited* home production, it may be assumed that the consumer will have to pay all the tariff imposition, much the same as though it were all imported. In cases where the domestic production is large, but yet has direct competition with the imported article, the duty is virtually shared between the foreign producer and American consumer. Unless an increase of revenue is imperatively demanded, all articles incapable of domestic production should be upon the free list. The non-dutiable schedule should also be extended to include articles of limited domestic production that are incapable of becoming unlimited. With any practical or probable unlimited home production, the interests of the domestic consumer will be safe. Any special *grades* of a general article, as of wool, that cannot be produced at home, should come in free.

If politics could be eliminated, it seems probable that in the light of sound economic principles, as briefly outlined above, an impartial tariff might be devised which, while not fully meeting sectional views and demands, would, on an average, be just to all.

The fact should not be overlooked that the tariff is not responsible for the decline of special industries where there has been a change in natural conditions. For instance, the production and working of iron and steel have decayed in some sections, not on account of the tariff, but because certain localities, as Pittsburg and Birmingham, are able to bring ore and coal together at lower rates than are possible elsewhere. Business will go to the cheapest producing points, and nothing can prevent it. As before suggested, all tariffs are obstructions, abstractly considered. If, in the future, national interests become broadened and universalized, the natural tendency will be toward a freer international exchange of products.

THE MODERN CORPORATION.

"*While they are subject to abuses, they are great forces in production, and have their place in the economic functions of society.*"

"*Corporations cannot commit treason, nor be outlawed, nor excommunicate, for they have no souls.*"

SIR EDWARD COKE.

XX.

THE MODERN CORPORATION.

The corporation has been a mighty instrumentality in the evolution of modern social conditions. In the steady growth and diffusion of the peaceful arts and industries among men, especially in the accomplishments of the last few decades, its force has been paramount. We can hardly conceive of the universal paralysis that would touch every phase of modern social life, if we were suddenly thrown back to a condition of absolute dependence upon personal units.

As the relation of the corporation to the shareholder is touched upon elsewhere, only its general relations to society, in the light of Natural Law, will here be considered.

The merits and demerits of the present application of the corporate principle are the subject of much popular discussion; and, here as elsewhere, in order to reach the bed-rock of logical truth we must discriminate between that which is normal and its prevalent abuses.

Corporate operation is ubiquitous. Any effort to live independent of its aid would at once result in the most primitive conditions. It is an agency through which human accomplishment is not only mechanically, but vitally multiplied. It builds and operates our railroads, telegraphs, steamships, and factories, develops our mines, transports our persons and property, manufactures our goods, and gives employment to both labor and capital. It is peculiar to a high order of social and moral development, and is not found to any extent elsewhere. Natural Law, as embodied in organization, is responsible for its existence and importance. It takes small units, and builds them up into those which are much greater.

Organization, as manifested on every plane, is characterized by force and utility. It is omnipresent in the physical world, erecting and cementing unities out of the most diverse materials.

Those nations and peoples who possess the genius for organization, and understand its power, are distinguished for the number and variety of their corporations. Such were the ancient Romans, and in Rome corporate organizations had their early development. The corporation of to-day, in Europe and America, is a later and broader utilization of the principle that was found useful by Rome when she was mistress of the world.

In general, corporations are divided into several classes, embracing the municipal, religious, educational, eleemosynary, and those of a commercial or business nature. As the latter compose the department under consideration, we shall confine our attention to them.

Business corporations are creations of the State, formed for the prosecution of enterprises which cannot be carried on so efficiently by individuals. Their object is the enhancement of the comfort and welfare of the whole people; but there was a time, in England, when royal charters conferred special and exclusive privileges. They are creations that have rights and obligations of their own, which are unlike those of their individual corporators.

A corporation is not like a single great capitalist. Its shareholders are usually people of moderate means, and the organization becomes powerful only through the aggregation of small sums representing shares. The office of this institution is to take this capital and wield it through chosen executive agents of unusual qualification, and thus increase its power and utility. If rightly handled, it is like a labor-saving machine to the shareholder. It is not merely a concentrated agency, to hire labor and sell product, but its office is to take crude and inexpert labor and capital, unitize them, and through skilful processes transmute them into new forms which, because more useful to society, will bring pecuniary reward. The corporation is not merely a great machine to be operated, but an economic, social, and moral force, so virile that it may advance as a pioneer far beyond the confines attainable to individual prospectors.

An English lawyer logically demonstrated that corporations "have no soul." He affirmed that none but God could create souls, and that corporations were creations of the king. This is often quoted as an inference that they are naturally hard and selfish. But the fact that they are soulless, in some ways gives them peculiar power and efficiency. Their impersonal quality frees them from individual weakness and idiosyncrasy. They are giants, reduced to order and put within boundaries through the sovereign power of the State.

The distinguishing corporate characteristic is perpetuity. The right of succession conferred by the State gives it great advantages over the individual. His operations close with his life, but this complex theoretical personage lives on. Officers and stockholders may die, or go into bankruptcy, but its activities are unceasing. However, there are three possible ways in which it may come to an end; viz., by the death of all its members without successors, which is extremely improbable; by a voluntary surrender of its charter; or by the repeal of its charter by the State. The latter, however, never takes place, except it be so provided in the charter, or in consequence of its violation, that being regarded as a contract between the State and the corporators. The corporation may make its own laws, provided they do not conflict with the general laws of the State. Except in the line of its own specific purpose, it is not as free as the individual. It must follow the prescribed path, while he may make any contract which is not actually unlawful in its nature. Its creator marks out its limits, and gives it a name under which persons act in specified ways for definite purposes.

It would be impossible for individuals to carry on great modern enterprises; but even if they had the ability, every thing would have to be closed up, or disposed of, with the settlement of their estates. The super-personal power of corporate enterprise alone renders higher social evolution possible.

Under our system of government, each State constitutes the sovereign power which creates and regulates the corporations which are located within its limits. There is, therefore, no uniformity in the powers, privileges, and limitations contained in the charters of American corporations. In some States the provisions for corporate establishment are general and simple, so that with little formality the necessary papers can be placed on file with the proper authorities and a new corporation brought into existence. The provisions bearing upon the relations of the corporation with its shareholders are also very unlike. Some require a full contribution of the par value of the shares, while others leave all such details without restriction. On account of these differences, corporations are often formed and chartered in States other than those where their business operations are mainly located. For railroads and other *quasi* public corporations, special legislative acts are required. The general tendency, however, is to substitute general laws and regulations for special acts, in the creation of corporations. Any peculiar or unusual chartered privileges have an exclusive and monopolistic aspect which is unpopular, and not in accord with democratic principles. Special legislation is rapidly giving way to that which is general, thereby placing the whole community on an equal footing.

The rapid increase in the number and variety of corporations, and their growing power, are suspiciously regarded by public sentiment. Here as elsewhere, it is easy to confound abuses with the system, and to overlook great usefulness and adaptation. We vastly overrate their power for harm, even if they have harmful motives. Their prosperity, as well as that of their corporators, is bound up in that of the body politic. As producers they are entirely dependent on demand, and can oblige no one to purchase their products, unless he may think it for his interest. As purchasers of labor or material, no one is obliged to sell to them except of his own free will. Even if the managing power of a corporation had savage instincts, it is securely caged in its outside relations by the natural principles of supply, demand, and competition, which are stronger than iron bars. While the public is therefore secure, stockholders are not always so well protected. They are behind the bars, and sometimes need to be saved from their "friends."

The popular impression of a corporation is that it is almost necessarily rich and selfish. But its peculiar function is not well understood. It is an economic industrial force, created for a specific purpose. Its executive management is a moral trusteeship for the carrying out of particular enterprise. It is not at liberty to divert the property of the stockholder into any channel outside of the special objects of the organization. Individuals should be benevolent, but it is a delicate ethical question whether or not corporate managers should be liberal with money which is not their own. It is entrusted to them for a specific purpose only. Benevolence is a function belonging to individuals, but not to a corporation, unless such an exercise is included in its charter. But every corporation should be humane, liberal, and moral, within the province of its recognized business. Individuals who compose a corporation, in their private capacity, should be altruistic toward all humanity; but the organization, as a unit, has had its limited scope definitely marked out at the time of its creation.

As a rule, the richest and most extensive corporations are composed of a large number of small stockholders. This is the case to an unappreciated extent with railroads, savings banks, and loan associations. Even the national banks, which in the eyes of many represent a small coterie of the rich, are found upon examination to be aggregations of capital generally owned in small amounts and widely scattered. Unfounded prejudice against legitimate wealth is often carried so far that it almost seems as though honest thrift and industry were something not very creditable. But the savings bank is an index, not only of thrift, but of education, refinement, and philanthropy.

The existing prejudice against corporations is not due to any inherent fault or lack of usefulness in the principle, but to the prevailing unfaithfulness among corporate managers. It lies in personal character — or rather the lack of it — and not in the system.

The primary movement in the establishment of new forces in material civilization is in the direction of temporary monopoly. But this is only a process. Its working is seen in the patent laws of all civilized nations. The secondary and permanent tendency is diffusive. There must be a gathered energy in the beginning to project new agencies into wide distribution. Meritorious inventions often fail to come into broad application because of the weakness of the centralized agency at their fountain head.

THE ABUSES OF CORPORATE MANAGEMENT.

"*Justice, sir, is the great interest of men on earth.*"
WEBSTER.

"*All power is a trust, and we are accountable for its exercise.*"
DISRAELI.

"*Private credit is wealth; public honor is security.*"
LETTERS OF JUNIUS.

"*No legacy is so rich as honesty.*"
ALL'S WELL THAT ENDS WELL.

"*An honest man's the noblest work of God.*"
POPE.

"*Render therefore to all their dues.*"
ROMANS xiii. 7.

"*Confidence is a plant of slow growth in an aged bosom.*"
PITT.

(*The Stockholder's Soliloquy.*)
"*Though this may be play to you,
'Tis death to us.*"
FABLE.

XXI.

THE ABUSES OF CORPORATE MANAGEMENT.

DIRECTORIAL unscrupulousness is a dark cloud in the social and moral horizon. It blunts public honesty, drags down pure ideals, chills wholesome enterprise, and furnishes a plausible excuse for socialistic and anarchic agitation. But it is no more a part of the normal corporation than are the barnacles a part of a graceful yacht. It is not an inherent part of the "social system," but a deadly upas, whose roots, trunk, and branches are all in blighted personal character.

It is often asked: Why is business so dull, good stocks so low, and money piled up in banks, instead of filling the channels of business? People scan the financial horizon, and assign almost every other reason for these conditions than the true one. Various theories place the fault in the tariff, the administration, too much silver, too little silver, surplus legislation, or needed legislation. But the true reason is *too little honesty*. General confidence in the integrity of the average manager and director has been badly shaken. The "lambs have been shorn," and "the goose killed that laid the golden egg," and others are not forthcoming.

The directorial board of a corporation, who are theoretically its servants and guardians, become its dictators and consumers. But the demoralization is general, rather than special; and it is to be feared that many who are not now officials would, under similar circumstances, do much the same. The manager is only a sort of exponent of the prevailing ethical standard.

Directorial abuses are not only common, but subtle, plausible, and insinuating, so as to obscure and almost eclipse axiomatic moral principles, which are older than the Decalogue. The public conscience is so accustomed to directorial manipulation, and skilful and prolific ingenuity on the part of officials, more especially those of the average railroad, that as a matter of course they are almost expected. To be on the "inside" is often as good as a fortune assured. Unscrupulous management is regarded only as "shrewd financiering," and even as "brilliant," so long as it escapes technical and legal cognizance and punishment. Instead of earnest condemnation from the public press, it often calls out criticism only of a flippant or facetious character. Its direct consequences may be seen in great congested, unearned fortunes, in a lax public conscience, in the universal distrust with which the world regards the average American railway management, and in the transformation of a legitimate stock-investment business into one of a gambling character. It furnishes the text and vantage-ground of every anarchist, socialist, and would-be destroyer of our present social order; and so far as legitimate investment is concerned, it has put it in close limitations. It seems strange that both legislation and public opinion have so lightly regarded these great commercial evils, and, while perfectly aware of their magnitude, have taken no earnest measures for their abatement. The direct financial sufferers are the hundreds of thousands of shareholders [1] and owners of other securities, who have furnished the nine or ten billions of dollars which have created the great arteries of American commerce, and without which the material resources of the nation would be but infantile in comparison with the present reality. The great majority of shareholders have no practical way of making their

---

[1] Their numbers range from a few hundred in some lines up to fifteen or twenty thousand each in a few of the great Western systems.

THE ABUSES OF CORPORATE MANAGEMENT. 249

influence felt in the conduct of the business of their several corporations, and are largely the victims of a false system which is strongly intrenched and of rapid growth.

With all due recognition of *numerous and honorable exceptions* in the application of these strictures, the fact remains that the average railway management is autocratic, irresponsible, and often definitely dishonest in its relations towards its shareholders. Investors are made to bleed at every pore, while their *pseudo* trustees, by means of earlier and superior information, fatten equally well on corporate adversity or prosperity. Between the triple combination of official dishonesty, the shackles of the interstate commerce law, and hostile State legislation, the interests of the shareholder are ground to powder.

Let us definitely note some of the ways in which the investor suffers from directorial chicanery.

First. By manipulation in the stock market, through combination or conspiracy among the managers, to the unfair disadvantage of the other proprietors.

Second. By withholding regular reports, statements, and information, for personal advantage, the injustice of which the common law does not take cognizance of.

Third. By complicated systems of book-keeping, which, though not technically fraudulent, are misleading and deceptive.

Fourth. By personal interest in other railways or corporations, with which consolidations, the absorption of branch lines, leases, or special running arrangements, are made, *to the detriment of common stockholders.*

Fifth. By official interest in railway "construction companies" and fast freight lines, for personal advantage, to the disadvantage of the corporation.

Sixth. By commissions and profits on, or an interest in, purchases or sales of material for the company.

If the early and peculiar knowledge incidental to the

position of the management be used as a basis for manipulation, to the detriment of the stockholder, it is as truly *stealing*, from a moral point of view, as it would be in case a portion of the rolling stock or track were purloined. Immaterial possessions are as intrinsic, and as truly *property*, as are those assets which can be seen and handled. The public conscience should be educated up to the point of calling these actions by their right names; and the lack of such explicitness has prompted proceedings, many of which are criminal in character, and which should be so classed in law and fact.

After this catalogue of managerial short-comings, some of our well-meaning nationalistic enthusiasts will doubtless respond: "Yes, the abuses are heinous and therefore the *Government* should acquire and operate these corporate properties — especially the railroads." Who is "the Government" practically? The politicians of the dominant party. Nationalization would add to the present array of evils the still more formidable corruption which inheres in political partisanship and "bossism." It might result in fastening on to the great arteries of commerce an enlarged national Tammany, the probable results of which may be faintly imagined.

But though a great and general reform in the public *morale* is the important thing needed, it may be possible for legislation, applied from without, to administer certain antidotes that are well worth a trial. Some of them may be outlined as follows: —

First. The compulsory issuing of monthly reports in a uniform manner, and after a prescribed formula, the correctness of which should be affirmed by the oath of one or more directors, adding thereto such explanatory matter as the management might deem necessary.

Second. A periodical audit by outside governmental examiners or professional accountants, duly qualified and sworn for this special service, on some plan similar to that used in the case of the national banks.

Third. Construe as bribery the receiving of any commissions or presents by any auditor, purchasing agent, or official, which are given because of his official position.

Fourth. That it shall be illegal, with heavy penalties, for any corporate official or manager to buy the stock of his company, except for actual investment; nor shall he sell the same unless he be the actual owner of the amount sold, and make a delivery of it; and he shall neither buy nor sell except after prescribed public notice.

Fifth. Require affidavits at stated intervals from each official that he has made no speculative sales or purchases, *indirectly*, of the stock of his corporation, and that he has no interest in any such transactions made by or through any other persons.

It may be objected that the last two proposed measures do not harmonize with the principles of *laissez faire;* but a sound political economy teaches that individual freedom must give way to collective freedom, and that the will of society is paramount to personal will. Managers would be restricted by such a plan only in a single direction, which is the vital point to be guarded, but elsewhere they would have perfect liberty. If the adoption of the last two measures were found impracticable, the single requirement of monthly sworn statements would strike a telling blow at the abuses of official control. Adequate salaries should be paid for all official services actually rendered, which would remove every plausible excuse on the part of the management for any predatory invasion of the shareholders' interests. It is time that the most necessary and honorable business of building and operating a railroad should be otherwise looked upon than as a "scheme," or even a game in which the management play with loaded dice.

The systematic wrecking and reorganizing of railways,

in which the interests of the shareholder are uniformly sacrificed, forms, perhaps, the darkest stain upon American commercial honor.

Practically, the average shareholder in a great railway corporation has no influence whatever in shaping its policy, even if he enact the farce of voting by self or proxy at its annual election. Is there, then, any remedy for the abuses which have been so harmful to the interests of investors, and which so seriously menace their future? It seems probable that a thoroughly organized working association of investors and stockholders in each important city might prove to be an exceedingly useful institution. Such organizations, by the employment of able committees on legal prosecution, legislation, economic literature, and in any other needed departments, might successfully grapple with abuses against which the individual is powerless. While advancing the interest of its own members, it also could greatly aid the general public in rooting out directorial manipulation and mismanagement, and in an important degree promote needed economic legislation, and also lend its aid in securing the repeal of that already in force which is harmful and superfluous. The moral as well as the legal power which such organizations might wield is very important.

Looked at from the outside, every railway or other business corporation is a unit. This fact is realized in all its external relations, whether with the general public, with legislation, or in its commercial transactions. Like the human body, though it has many members, it has but one will, one head, one voice. But though a unit as viewed from without, its internal relations are many and complex. The usurpations of the management, who entirely control and regulate its outside relations, often entirely defeat the course intended from within. Stockholders should therefore be more vigilant in guarding their own interests.

Through combined effort they can make their influence felt, instead of passively perpetuating mismanagement, through the use of the conventional corporate machinery. Any well ordered efforts in this direction would not only conserve interests that were directly represented, but also have a wholesome moral influence in a broader scope.

THE EVOLUTION OF THE RAILROAD.

"*No pent-up Utica contracts your powers,
But the whole boundless continent is yours.*"
                                                    SEWALL.

"*Facility of communication in social, commercial, and political intercourse is a distinguishing index of civilization.*"

"*Let observation, with extensive view,
Survey mankind from China to Peru.*"
                                                    SAMUEL JOHNSON.

XXII.

THE EVOLUTION OF THE RAILROAD.

PROGRESS toward higher civilization and social development is hastened by the growing perfection of the means of communication and transportation. The body politic, like the human organism, must have its vital currents; and their circulatory processes are carried forward by means of arteries and veins. The roads of a country provide for its pulsations of living activity, and their superiority is an index of its progress, not only in commerce, but in art, science, and literature. Barbarism is universally characterized by the lack of any adequate facilities for travel and commercial intercourse.

It is therefore in accord with Natural Law that the presence or absence of adequate roads indicates a dividing line between two diverse conditions of society. When the Roman Empire was at the height of its power and grandeur, it was distinguished for its roads, and all led to Rome. Portions of the famous Appian Way, built three hundred years before the Christian era, still remain. It was over three hundred miles in length, spacious, and smoothly paved with hewn stone blocks, laid in cement. Numerous other roads, equal in character to our best city streets, diverged from Rome for thousands of miles, to the most distant parts of the empire. Mountains of rock were tunnelled, and rivers and ravines were spanned by massive stone bridges, over which her invincible legions could march without interruption, while she was mistress of the world. These great works were so substantial that fragments of them still remain, notwithstanding the disintegrating influences of the frosts and floods of twenty centuries. In contrast, the feudal age of comparative barbarism was destitute of highways, and had no facilities for communication. The baron of old England, or on the Rhine, who ruled the adjacent region, perched his castle on inaccessible heights. He built no roads or bridges, for communication was not desired. Wheeled vehicles, except a few of the rudest sort, were unknown, and all travelling was done on foot or on horseback, through fields, forests, and streams. Then there could be no social or mental progress, no commerce, and no reciprocal activity. Industrial development was impossible for lack of channels. By Natural Law, friction produces heat; so inter-communication excites mental activity, and stimulates art, science, and invention. Nothing has so contributed to dispel the lethargy of ages, and to quicken the current of investigation, as the utilization of steam and electricity.

Even turnpikes were not constructed in England until the early part of the last century, and the first English canal was dug as late as 1760. The yearly movement of merchandise on all the through land routes of the world a century ago, would not equal that of one of our great trunk lines of the present time. Long-distance transportation by land, except for the most concentrated and valuable products, is entirely a thing of the present. We are more inclined to look at the present and the future, but a brief retrospect is often instructive. Not till 1833 was there a daily mail between London and Paris. The English postage on foreign letters was from twenty-eight to eighty-four cents, besides the foreign rates and ship charges to be paid by the receiver. On inland letters, at the same time, the postage was twenty cents per sheet. In our own country, up to 1845, inland rates were from six to twenty-five cents, according to distance. In 1851, a reduction was made to a uniform rate of three cents per half ounce. Not only modern civilization depends upon easy and rapid communication, but even free government, except on a small scale, could hardly exist without it. Union of sentiment is indispensable to its continuance, and modern facilities of intercourse alone can secure it. The people of a vast territory, like our own, are more thoroughly assimilated and unified than was possible a century ago with those of a single State. The far away provincial towns feel the metropolitan heart-throbs.

We soon become accustomed to modern facilities, accept them as a matter of course, and regard their usefulness with indifference. Not only so, but we become exacting and almost unreasonable in our demands upon them. The prairie farmer, who perhaps formerly used his corn as fuel for lack of transportation and a market, soon forgets his experience, and is dissatisfied with his present advantages. The railroad, which has doubled the value of his farm and products, and for the completion of which he ardently longed, soon becomes to him an offensive monopoly.

A hundred years ago it cost three dollars to transport a barrel of flour a hundred miles; and salt which was a cent a pound at a seaport, often cost six cents at an inland market.

A part of the price of all products is made up of their cost of carriage from the place where they were grown or manufactured. Often a slight decrease in transportation charges creates new business, and enlarges that before established a hundred-fold, rendering necessary a large increase in the labor required.

When railroads were in their infancy, it was assumed that they would be public highways, and that every shipper would use his own cars, or trains, paying the company a toll for the use of their track. As business increased, it was soon found that such a plan was utterly impracticable.

The present clamor for restrictive legislation is perhaps a remnant of this antiquated idea, and much of that proposed is no more practical. A railroad is not merely an improved public highway, but a great and complicated transporting machine, requiring the highest order of ability for its successful operation. We are mainly considering public interests as related to railroads, but will briefly look at those of investors. We have seen that it is natural that, as the interval between the investor and investment increases, the dangers from waste and mismanagement increase in like proportion. On this point Mr. John B. Jervis, in his able work on railway property, says: "This kind of investment is not well suited in general to small proprietors so situated that they can exercise no control, and who are exposed to the danger of having their property managed by unfaithful men, who seek to make the institution subservient to their interest, rather than to that of the proprietors."

The remarkable movement towards consolidation, which has taken place during the last thirty years, deserves brief attention.. Popular sentiment is distrustful of growing aggregations of capital and power, and some look upon them as an evil, or even as a menace to our institutions. The fact that consolidation is not only caused by Natural Law, but is also ruled by it, is entirely overlooked. If the process went on, until there were only one gigantic system in the whole country, it would still be subservient to the imperial edicts of supply and demand. If it made an effort to impose artificial rates, or those that were even a little above the normal, then in a *greater proportion* demand would fall off and business and profits decrease. If a normal rate were restored, demand for service would be so much enhanced that financial gain would result. Supply and demand perform their office as quickly and surely as does the "governor" of a steam engine.

The earliest railroad charters were for short independent

THE EVOLUTION OF THE RAILROAD. 261

lines. In England the earlier railways averaged only fifteen miles in length. In 1847, five thousand miles were owned by several hundred different companies. In 1872, thirteen thousand miles were nearly all owned by twelve companies. This tendency has been nearly as marked in this country. As a single instance, that part of the New York Central line between the Hudson River and Lake Erie, originally belonged to sixteen different companies. During the last few years the development has been, not merely into longer lines, but into great systems. Many of these now embrace from two thousand to six thousand miles of road, and form arteries through which commercial currents flow, giving life to great domains, each larger than some of the entire kingdoms of the Old World. What is the cause of this general and rapid consolidation? what are its tendencies? and what will be the results? It has taken place not by chance, nor because of any local or temporary reasons, but in obedience to the pressure and behests of unvarying Natural Law. The natural demand for decreasing rates of transportation, together with competition, have made it indispensable. It is a case of the "survival of the fittest," and of a development of the lower into the higher. In no other way could such remarkable reductions in rates and vast increase of business have been realized. Under no other plan would such a degree of perfection in appliances and rapidity of service be possible. Modern convenience, comfort, and luxury are the results of the law of combination and consolidation. Contrast the present passenger service with that of thirty years ago. A passenger leaving New York for Chicago not only paid a much higher fare, but had to change at the end of each separate short line, and was as often compelled to stand in line to get baggage rechecked and reloaded, subject to frequent lack of connection, long hours of waiting, and other numerous discomforts. One consolidated system of a thousand miles in length can render to the public a service which is immeasurably superior in luxury, cheapness, speed, and safety, to that which would be possible with any half-dozen distinct corporations.

The special and unprecedented national legislation known as the Interstate Commerce Law has now been in force for several years; and, on the whole, its working seems to have been detrimental to the owners of railroad property with little compensating advantage to the general public. This law is based upon that clause in the Constitution which includes, among the duties of Congress, the regulation of commerce between the States. The plain intention of the framers of the Constitution was to forever prevent, by any State, the erection of any customs tariffs, so that State lines should be no obstruction to the free currents of commerce. The idea of regulating the market price for carrying freight or passengers probably never entered their minds. The basis for a national legislative interference with the legitimate, free, competitive business of common carriers, therefore seems strained and unnatural. As the courts have sanctioned such regulation, however, their decision must be accepted. It is undoubtedly true that the peculiar public sentiment which inspired this law was mainly prompted by directorial abuses; but the aim was wide of the mark, for when the charge was fired, the game that came down was the shareholder.

The object of government and legislation is not to destroy value, but to protect and conserve it. The Interstate law, through the long and short haul clause, and the prohibition of pooling, accomplishes the former. The right to arbitrarily regulate the rates for freight or passenger service, and thus place not only the earnings but the absolute capital — which derives its entire value from expected earnings — at the mercy of a possible unintelligent majority of Congress, even through a political commission, is the most radical and questionable legislative experiment of our national history. It is an assumption of legal authority which lacks a proper moral basis. The stiff unscientific hand of legislation undertakes to regulate prices which inherently have in them the elements of self-regulation. Objection to this will be made on the ground that railways are *quasi*-public institutions, and that by their charters they have been granted the right of eminent domain; all of which is true, but the sole reason for granting that right was the public convenience, and the shareholder paid full value for every square foot of ground taken or damaged.

Two ever-present natural principles regulate the price of railway service, as of every other salable commodity: (1) direct competition, and (2) the invariable fact that demand falls off the moment that prices are placed above the normal point. The latter principle, even in the entire absence of the former, hedges in the strongest corporation. The greatest profit in the long run is always realized at fair rather than at exorbitant rates, popular opinion to the contrary notwithstanding.

As a proof, witness the steady decline in actual rates for railway service, as they kept pace with the normal — or those prices for which, owing to continually improving appliances, it could be rendered — before there was any legislation on the subject. A reduction of ten per cent in rates often brings an increase of twenty-five or even fifty per cent in business.

Govermental regulation sounds well; but the real, not the ideal government, may consist of a bare majority of unexpert and unconscientious politicians who happen to compose a given Congress. The material and financial conditions of no two of the hundreds of American railways are exactly alike; and while the natural and elastic law of supply and demand can adjust itself to them all, it is impossible for legislative law to accomplish this, as it has no flexibility.

The Interstate long and short haul clause is economically irrational, and destructive of investors' interests, because all price making, except in the case of *natural* monopolies which receive charters, is beyond the province of legislation.

Can any other law be named to which the enforcement, or permission to ignore, is left optional in each case with a board of extra-judicial commissioners? The law vests in a few men a power which is autocratic. The duty of the judiciary is to interpret and to enforce law; but this commission has the power to settle questions involving millions, as mere problems of expediency, in which no principle of right or wrong is involved. This Interstate law has already caused a shrinkage of hundreds of millions in the actual investments of American shareholders, and all without any corresponding advantage to shippers. The shipper, during a general business depression caused by the shrinkage in railway values, suffers in common with all other classes. If, however, shippers were benefited, it would be at the expense of justice. That section of the law which prohibits pooling naturally forces further and greater consolidation, as the only alternative to general bankruptcy. Business can only flourish under free conditions, and the true province of legislation is to enforce contracts which have been voluntarily entered into.

But there is a section of evident utility in the Interstate law, and that is the one which prohibits discrimination under like circumstances and conditions. This section, unlike the two before examined, is founded on a moral principle. In the main, the Interstate law is the embodied expression of an unreasonable prejudice against vested interests as represented in the railway corporations of the nation. While directorial manipulation has greatly intensified and made ostensibly reasonable this feeling, it is at the same time blind in its aim, and unjustifiable in its antagonism to the rights and privileges of the shareholder.

THE EVOLUTION OF THE RAILROAD. 265

Abnormally cheap long hauls, under free conditions, created a vast business, which is destroyed by the *pro rata* shackles of the "long and short haul clause." In many cases, if the roads are forbidden to do it at a somewhat less price, proportionately, than is charged for shorter hauls, the business is lost. A railway must maintain its facilities and fixed charges in either event, and therefore cheap long hauls are almost so much clear gain to the ordinary revenues of the road. With this source of income cut off, the deficiency in the long run must be made up by higher way rates than were necessary before the freedom of railway transportation was taken away. The Canadian Pacific line, being free from the incubus of artificial restraint, is able to command an immense through business that would naturally be done by American lines, which fact in itself forms a striking commentary on the wisdom of "long and short haul" repression.

Rates for railway service in the United States when compared with those demanded in Europe are found to be astonishingly low, notwithstanding the fact that employees' wages are more than double the European average. The rate charged per ton per mile by the great trunk lines running from Chicago to the Atlantic seaboard, as shown by statistics, are a little less than one quarter of the average rates of the year 1865. On the great systems west, northwest, and southwest of Chicago the charges have been reduced nearly in the same proportion.

Restrictive railway legislation, as expressed in the recent enactments of several Western States, is unjust to the shareholder who has investments in the lines of those localities. Not content with the appointment of commissioners whose duties shall be to fairly and carefully conserve the interests of both the public and the roads, they construct ready-made tariffs, fix unreasonable maximum prices, and shackle what little commercial freedom has been spared by national enactments.

The result of the repressive policy, national and State, while disastrous to the dividend-earning capacity of the railways, at the same time will prove morally and financially detrimental to the whole country. "When one member suffers all suffer." The track, road-bed, and rolling-stock of bankrupt, insolvent, and non-dividend earning lines necessarily deteriorate, and the public service becomes indifferent. Railway enterprise, responsibility, and reliability will become impaired, unless the restrictive policy now so confidently pursued is relaxed in an important degree.

The downward tendency in rates for the transportation of freight, before noted, which was steady and continuous prior to general legislative restriction, was in obedience to the principles of supply and demand. Greatly improved appliances, wielded by wider and more thoroughly organized control, cause the normal rates for service to decline, and natural principles are ceaselessly pressing actual rates into conformity. A normal rate is that point above which demand falls off so rapidly that profits diminish, and below which even a great increase of business would lessen them. The problem with railroad management is, therefore, to make the nearest possible approximation to it. It varies with every road, and with every different class of freight, and is a very complicated question, and one entirely beyond the province of legal enactment. How much each particular variety of freight will bear, without in any way hindering its greatest possible increase and development, is a very delicate problem, and must be solved with great care. Legislation is futile, not only because price-making is outside of its province, but because no two roads are alike in business location, cost of maintenance, character of traffic, and many other conditions. There is the same variety in these as in individual enterprises. To a great extent, rates fix themselves, and the power of the management, in this respect, is greatly overrated. Suppose two or more competing lines enter into a "cast-iron" agreement to fix rates that are somewhat above the normal. How soon shipments fall off, rival routes or water transportation compete, markets are disturbed, and speedily the "cast-iron" crumbles. Take the extreme case of a road that has no possible competition. If even such a road attempts to impose artificial rates, business is hampered, settlement of tributary territory discouraged, manufactures excluded, and profits actually diminished. All railroad men have not yet become aware of these laws, but they are rapidly learning them, and also that a broad and liberal policy is the most profitable. In no other kind of business is the old familiar principle of "large sales and small profits" so applicable and profitable as in railroad transportation. The reason for this is that a large part of the expenses consist of "fixed charges," which are unchanged, whether the traffic be large or small. Outside of these, expenses increase much more slowly than the amount of business. An increase of twenty-five per cent in general expenses might be sufficient for a business one hundred per cent greater. A system of five thousand miles probably would not cost half as much to operate as it did formerly when made up of a dozen distinct corporations. It has one board of management, instead of many; unity of purpose in place of diversity; single and thorough organization instead of inharmonious variety. The friction of one large wheel is much less than that of many small ones, and its power and momentum are vastly greater. Consolidation should be considered as the greatest labor and expense saving process of the age. Why should "reformers" make such efforts to excite popular prejudice against consolidations? Is it simply because they are great? This is an age of grand things, and of wonderful privileges and benefits that are lightly appreciated. A sentimental cry of "monopoly" seems to be all that is necessary to arouse unthinking popular prejudice.

The greatest possible consolidation is hedged in on every side by the impregnable, though invisible, barriers of Natural Law. With constantly diminishing rates for service, and increasing safety, luxury, and rapidity, it has not yet been explained how the modern railroad can be a "menace," either to the citizen or the government. Granted that sharp practice, stock watering, and many abuses exist, both in the construction and operation of these great thoroughfares, the systems are here, and are to remain, and the public gets the advantage. Abuses are incidental to every enterprise, no matter how meritorious; and this will be the rule as long as the element of selfishness is dominant in human nature. Statistics show that in a majority of cases the original stockholders sunk the money invested, and that the lines were afterwards and are still being operated by other and different proprietors, who purchased the assets at a nominal price. In general, no other investments pay so small a rate of interest as those in railroad property. Stock watering is indefensible as a system; but a candid view will show that, in some cases, it is only a "marking up" of nominal value to correspond with what has taken place in actual value. The enhancement of market and taxable values of terminal facilities and other kinds of property and improvements, is often considerable in a series of years. For instance: If the taxable and the salable value of a road have increased fifty per cent in ten years, is an increase of the stock by which it is represented in the same proportion in any way illegitimate? While this is the popular impression, there seems to be no valid reason why railroad property should be exceptional in this respect.

Another prevalent fallacy is that stock watering necessarily results in a higher tariff. We have already shown that rates are made by causes entirely different. If the nominal amount of the stock of any road were quadrupled, or reduced in the same ratio, its material property still remains unchanged. Its earning capacities, surroundings, facilities, and opportunities are neither increased nor diminished. The normal rates at which business and profits are at the maximum continue as before.

Wherever there is dishonest and extravagant management, the investor suffers; but the public escapes, except indirectly. If unreasonable popular prejudice were gratified to the extent that, by unfriendly legislation, these great corporations could be crushed, it would be found that scores of small owners would be ruined as often as one "millionaire." A large majority of the stock and bonds of these corporations is widely scattered among thousands of small holders, including even many widows and orphans. Hundreds of millions of dollars have been lost by investors, the benefit of which is now being realized by the public. The commercial importance of these far-reaching systems is even excelled by their moral and political value in unifying all our diverse sections and interests.

The highest order of executive talent is required for their successful management. The chimerical plan that the control of these vast interests should be assumed by the general government, and so become the sport of politicians, to be fought over every four years, is unworthy of serious consideration.

Cheap and rapid transportation has created new commercial centres, and millions of worthless acres have not only been transformed into productive farms, but have practically been moved a thousand miles nearer to market. The "long-haul" business is entirely a thing of recent times. The food products of the great trans-Mississippi region are found in the European markets, through the practical annihilation of distance by the power of modern transportation.

Art, science, and literature have all felt the quickening influence of this movement. Nothing since the invention of the printing press has so accelerated thought and investigation. With the aid of steam and electricity a nation becomes a neighborhood, and the pulsations of news, politics, morals, and religion are felt to the extremities. Mind attains increased preponderance over matter, the natural way of advancement is opened, and a new *Renaissance* is ushered in. By Natural Law, physical, mental, and moral attainment depend upon man's grasp and utilization of the forces with which nature's storehouse is overflowing.

INDUSTRIAL EDUCATION.

"*Let a man have accurate perceptions. Let him, if he have hands, handle; if eyes, measure and discriminate; let him accept and hive every fact of chemistry, natural history, and economics; the more he has, the less is he willing to spare any one. Time is always bringing the occasions that disclose their value. Some wisdom comes out of every natural and innocent action. The application of means to ends insures victory and the songs of victory, not less in a farm or a shop than in the tactics of party or of war. One might find argument for optimism in the abundant flow of this saccharine element of pleasure in every suburb and extremity of the good world. Let a man keep the law — any law — and his way will be strewn with satisfactions.*"   EMERSON.

"*Stretch'd on the rack of a too easy chair,
And heard thy everlasting yawn confess
The pains and penalties of idleness.*"
POPE.

XXIII.

INDUSTRIAL EDUCATION.

We are living in a peculiar era. While Natural Law is unchangeable, its multiform applications are ever shifting. Old customs, conditions, and methods of thought are being superseded. The necessity for a readjustment is nowhere more marked than in the department of education. Former methods of training for active life require revision, almost revolution, in order that they may satisfy modern requirements. Through changed conditions and the antagonism of trade unions, the apprentice system, once so universal, is almost utterly extinct, and the rising generation is brought face to face with serious problems.

The conventional intellectual education of American youth is clearly inadequate to meet present social and economic demands. The regular professions, including the law, medicine, and theology — and even journalism and teaching — are already overcrowded and constantly becoming more so. American young men are too numerous to find occupation in the purely intellectual professions, and wider opportunities must be afforded. The prevalent idea, especially among the well-to-do classes, that their sons must employ their talents within this select and limited range, must be modified, else overcrowding, inefficiency, and idleness will be prevalent and disastrous.

Turning to other and more available pursuits, we find that their exercise and emoluments are rapidly slipping away from native-born youth and being grasped by those of foreign birth and training. A plain statement of the facts involves no racial prejudice or exclusiveness, but only a recognition of certain needed readjustments which are highly proper and equitable.

What shall the average American boy do? In what channel can his faculties be trained so that he shall be a producer, or a useful member of society, and also by means of such economic activity be able to gain a livelihood. Through a virtual monopoly of all the principal industrial trades and handicrafts by the unions, which are controlled and almost entirely composed of the foreign born, the American boy has come to be almost a superfluity. Opportunities for thoroughly learning any industrial occupation are well nigh lost to him. To further aggravate the difficulties of his position a prevailing false pride — shared not only by himself but often by his parents — causes him to look askance at manual employment.

While the American common school system, with its free and even compulsory intellectual training, has been a matter of just gratulation, and while it has filled an important place, it must still be admitted that its limitations are many and important. There has been some improvement in the general ideal of what true education — or educing — is; but it still remains that the schools of all grades deal almost entirely with objective statements rather than subjective mental unfoldment. Their principal aim seems to be the acquirement of a great volume of unrelated facts of limited practical significance, instead of a development of the latent capabilities. The popular idea of education is still, that it consists of knowing a great many things, whether or not these form any part of an all-around equipment for the duties of active life.

Conventional education is also greatly destitute of the ethical element. It cannot be denied that each church and religious denomination has been indifferent, if not averse, to any moral system, lest some theological bias might creep in which would be more favorable to the others than to itself. A regulation *curriculum* ethically colorless has been thought sufficient, the foundations for character and good citizenship being largely neglected. Laying aside all sectarian suspicion, it would not seem difficult to agree upon a moral discipline to which not even the smallest minority of a civilized community would make objection. All will admit that animalism and selfishness, which are so obtrusive in energetic youthful character, need to be guided and moulded into higher activities. Without cant or affectation the teacher could inculcate the manly and womanly virtues of temperance, honesty, charity, chastity, truthfulness, justice, order, frugality, industry, honor, cleanliness of mind and body, generosity, gratitude, parental respect, heroism, patriotism, reverence, self-respect, and manners, — all of which every citizen, be he Protestant, Catholic, Jew, or Agnostic, would heartily approve. Would not more of these abundantly compensate for a little less detail regarding the rivers of China, the mountains of Siberia, or the campaigns of Scipio? Is this all relevant in political economy? Vitally so, for State education has for its purpose the making of good citizens. General social and economic prosperity will depend upon the average youthful development of character.

Mere intellectual exercise only increases a force which without ethical regulation is mischievous and hazardous. If "the worship of the almighty dollar," and the dominion of a sensuous materialism, are ever to be made secondary to higher aspiration, the germs of such a growth must be planted in the fertile soil of plastic mind. The ideal society can only be composed of ideal men. The youth who is ethically weak is but a bundle of disorderly and unbalanced propensities, and mere intellectual equipment increases their energy. The great educational lack of the present time is in morality and industry. These must prevail as the rule, else the condition of the State itself will become precarious. A stream can rise no higher than its source, and collective character will not excel that of the average individual. A verbose knowledge of objective facts will in no degree compensate for a lack of the foundations of good citizenship.

When the average American boy "completes his education" in high school, academy, or college, what can he do for the world, either mentally or physically, that will be of service, and at the same time give him economic support? How can he stamp the regulation stock of learning which he has acquired into the coinage of society? If he be unable to do this it will soon decay from lack of exercise. As before noted, only a minority of college graduates can find places in the "professions," with journalism and teaching included.

But what of the much larger number who conclude their course of study with the common or high school, and then wish to enter the active duties of life? They find the available fields of activity all filled to repletion, and are unable to get a foot-hold. Pride and lack of opportunity then forbid them to *begin* to learn the rudiments of any industrial trade or handicraft. Their educational outfit has disqualified rather than fitted them to make any favorable start in a promising calling.

The demand for sales-people, clerks, office-help, cashiers, and copyists, having been largely filled by girls, and the industrial trades by foreign unionists, the American boy after the completion of his school course finds himself stranded. He has reached high-water mark and deterioration begins. He cannot utilize the kind of education he has received; and being unable to find occupation suited to his tastes and abilities, he soon drifts toward idleness and inefficiency. Even if of affluent family he is in reality much more helpless than his foreign-born associate who has been bred to manual effort. In addition to a false pride and a spurious social standard, the American boy has to contend with lack of opportunity and an education which is a misfit.

It is certain that manual training and trade schools, established and made efficient by the State, would greatly aid in the accomplishment of two most desirable objects. It would, in a measure, supply the missing education that has been lost through modern conditions and the decay of apprenticeship, and also greatly ennoble and dignify manual employment. It would furnish a potent remedial agency for the idleness and degeneration which are becoming so serious and prevalent.

Some degree of manual dexterity, and ability to use tools skilfully, is a valuable resource for every young man, even if he is to engage in a purely intellectual profession.

It must be understood that there is a wide distinction between simple manual training and the full acquisition of special trades. A practical adjunct of the former could be made at small expense to almost every common school in the land, while the teaching of specific trades would only be possible where large institutional plants or shops could be centralized. Manual training, even with a few tools, so develops industry and precision, and educates the hands and eyes, that it lays a general foundation for all trades, and there would be little difficulty in making it a factor in every common school course of instruction. Even brief manual cultivation would be of uniform profit, whatever might be the prospective avocation. It might be defined as athletics made useful, and at the same time reasonable in its intensity. Its reflex action upon the mind, while lightly appreciated, is of high importance. It directly cultivates and stimulates care, exactitude, promptness, celerity, proportion, and even honesty. Every physical process presumes a previous subjective plan and mental picture. A few years ago a gifted New York lawyer[1] wrote a good-sized volume entitled "Mechanics and Faith." In a most interesting and logical manner he traced out the correspondences and revelations, mental and spiritual, which a study of the principles of mechanics unfolds and indorses. Mechanical science, usually regarded as dry, is, in reality, rich in elements of beauty and even poetry. Applied mechanical principles through systematic instruction would do much to idealize their exercise and lift them from the realm of drudgery. To make by hand even so simple a thing as a symmetrical box is not only a physical, but a mental and even artistic accomplishment. The most common works of life are capable of redemption from the prosy domain of duty and toil, and of investment with grace and dignity from a change of standpoint. Through a possible educational treatment the plainest tasks can be idealized and made attractive.

When one considers the immense amount of wealth that is poured into the endowment funds of conventional colleges where men are turned out all of a regulation pattern, whose education is largely unavailable and often utterly useless, it seems like a partial misdirection of a great possibility. The splendid example of Colonel Auchmuty of New York City, who founded the great industrial school which bears his name, is worthy of the imitation of other wealthy gentlemen who are bestowing their millions in the endowment of the purely intellectual institutions, of which the country is already abundantly supplied. It is true that technical education is receiving more attention than heretofore, and that a limited number are availing themselves of its advantages; but it should become vastly more general, and instead of a very few institutions there should be scores. The methods and varieties of industrial effort are so numerous that there is ample room

[1] Charles Talbot Porter.

in this great field, both for the State and for many privately endowed institutions. Its principles are capable of indefinite application and expansion. It may extend from a simple annex to the common school house, where for an hour a day boys could learn the use of a few simple tools, up to the most elaborately equipped trade-shops and great technical institutions.

The average boy needs some useful outlet for his abounding physical activity. Often he has no taste for books, and study to him is perfunctory and mechanical. He is an untrained and ungoverned *force*, liable not only to useless but to harmful activities. He needs something to arouse his interest, develop his latent faculties, and to turn his overflowing energy into some useful and practical channel. An hour a day of manual training with its precision and calculation will do much to evolve true manliness and self-reliance. To create or construct something tangible tends to inspire character. The development of one set of human faculties, through related and reflex influences, invigorates all the others, therefore exercises should be frequently changed or alternated.

Statistics show that a large proportion of the criminal class begin in their downward career between the ages of fifteen and twenty-one. They start out to be honest; but idleness, which is the parent of all mischief, causes them gradually to drift into crime. They find no vacancies, either in business or the professions, and with expanding wants have no honorable means to supply them. Had they been educated to industrial dexterity they could be of use to themselves and to society.

Hon. Samuel B. Capen, for many years President of the Boston School Board, and an active philanthropist, who is much interested in the subject of industrial education, says: —

"That there is a special reason for trade-schools is made more apparent when we remember that, out of every one hundred boys that graduate from our grammar schools, only one per cent enter the ministry, one per cent become lawyers, one per cent physicians, five per cent business men, and ninety-two per cent get a livelihood by their hands. Are we doing all we ought for the ninety-two per cent?"

Another able writer and expert, in speaking of trade-schools, says: —

"The trade-unions in these cities are controlled by foreigners, who seek to confine their industries to men of their own nationalities. They not only refuse to teach an American boy a trade, but they combine to prevent him from getting employment after he has succeeded in learning it in a trade-school. This is a situation of affairs without parallel in any country in the world, and one which will not be tolerated in this country when once public opinion has been aroused to a full comprehension of it. It is surely not too much for the American people to say that their own sons shall not only be permitted to learn trades, but shall be permitted also to work at them after they have learned them. We advise any one who is desirous of seeing the kind of skilled workingmen that the American boy makes to visit Colonel Auchmuty's schools and look over a set of photographs of his graduates. He will find there a body of clear-browed, straight-eyed young fellows who will compare well with the graduates of our colleges. This is the stuff from which laborers are made who honor and dignify and elevate labor, not by agitating, but by being masters of their craft, faithful in performance, and willing to share its toil with all comers, fearing honest competition from no quarter. Such men are at once true American laborers and true American citizens of the highest type, and the educational system which evolves them is a national benefaction of incalculable value."

In seeking for remedial agencies for the vast amount of social and economic infelicity of the present time, there is nothing so promising, and which contains such grand possibilities, as industrial training. It should become as universal as the present intellectual courses of instruction. To co-educate the head and hands is advantageous for both. If every common school in the land could have an annex, used for the cultivation of manual dexterity, it would be a long step toward the elimination of prevailing sociological ills. Manual labor must be lifted and dignified by an admixture of the intellectual element. It can be rendered positively attractive by judicious idealization.

NATURAL LAW AND IDEALISM.

> "Because the soul is progressive, it never quite repeats itself, but in every act attempts the production of a new and fairer whole."
>
> EMERSON.

> "All nature is but art unknown to thee;
> All chance, direction which thou canst not see."
>
> POPE.

> "Up, my comrades! up and doing!
> Manhood's rugged play
> Still renewing, bravely hewing
> Through the world our way."
>
> WHITTIER.

> "That very law which moulds a tear,
> And bids it trickle from its source,
> That law preserves the earth a sphere,
> And guides the planets in their course."
>
> SAMUEL ROGERS.

> "Facts are stubborn things."
>
> SMOLLETT.

XXIV.

NATURAL LAW AND IDEALISM.

THE universal reign of law is the grand truth, which, if everywhere recognized, would transform the world. All human infelicity, whether physical, social, economic, moral, or spiritual, comes from a disregard or violation of the Established Order. Law will not and cannot bend to human caprice, for its lines are immutable. It is the final and infallible touch-stone which tests every opinion, institution, and system, and from its verdict there is no appeal. Its exact trial-balances and compensations put out of the question all cheats and short-cuts, while chance and even injustice are neutralized in the last analysis. It is a ubiquitous and righteous Judge, whose mandates can neither be dodged nor compromised with. The harvest will bear the likeness of the seed that was sown, whether in economics, morals, or any other realm of the mental economy.

We are largely blind to the universal supremacy of Law because we fail to recognize the positive relationship and interdependence of all things. Every event and principle is related to and modified by everything else. Each one of these invisible but unbreakable ties possesses a significance and conveys an influence. As no boundary is possible, except in the human consciousness, there can be no obstruction to their orderly vibrations.

The acceptance of this fundamental establishment naturally leads to the consideration of two sequential problems, which may be stated as follows: What is the nature of Law? and, How may we become intelligently certain of the direction of its leadings? If it be unrepealable, it is highly important to find whether or not it is uniformly good. A superficial glance at this vital question may make its answer appear doubtful. Take the supposed law of "the survival of the fittest," which seems to have a wide application, not only in the physical but in the social and economic realm. How can it be beneficent? It may at once be admitted that from a material point of view alone, this and other inherent tendencies appear adverse. Is that the correct standpoint? This directly suggests another question: Is man body or mind? If any one objects that such a query belongs to metaphysics, rather than to political economy, we reply that the science of economics is in the mind of man. Labor, capital, money, coinage, and tariffs are only external and resultant phenomena. Disconnected from their subjective relations they have no significance. Intrinsic political economy is written within man's constitution, while the things so designated are only their visible articulation. The outward manifestation is only the shadow of the internal substance. All veritable social science is therefore subjective, or in other words, metaphysical. For this reason its conventional treatment is like the play of Hamlet with the principal character missing.

But recurring to the nature of Law, we take the positive ground that it is as beneficent as it is universal. Only the standpoint which takes account of an all-comprehensive evolutionary trend will reveal this significant fact. Turning to first principles: If the grand purpose of creation be good, all its minor processes must be tending, even though indirectly, in the same direction. Law is only a comprehensive name for the orderly methods of the Creator. The supreme uniformity and reliability of phenomena prove that they are divine manifestations, and that only. An approximate human conformity to our highest interpretation of Law we call good; and such a lack of, or non-conformity as is below this standard we designate as evil. Only a higher and truer standpoint than the external and material will enable us wisely to interpret many forces of the physical and economic domain which seem destructive.

The so-called law of selfishness when viewed from its own plane seems to be Law, but from the altitude of unselfishness it is only relative immaturity. It appears to be Law from its being so general. The true beneficence of Law is found only in the breadth of its application; as, for instance, the seeming good of the individual lies in his own sole advantage, but a deeper view shows that his truer benefit is only contained in wider relationship. The individual good can only find its highest realization *within* the general good. His supremest development cannot take place by itself. The "fittest" gain that position only by being channels for the less fit, and the latter need the former for inspiration and example. The selfish rich man is not only socially, but individually unfit, for his apparent completeness is only superficial. Selfishness, faction, antagonism, envy, and avarice, though having a kind of regularity which makes them seem like laws, form no part of Law. The latter being uniformly beneficent lends its benediction just in the proportion that its methods are complied with.

Law is not a great, blind, mechanical force, crushing its violators and opponents, but an infinitely potent agency to be intelligently wielded and utilized. Its wonderful possibilities are placed at our service. They are like the mechanical forces of the screw or lever to the artisan, but extend in all directions and through all relations.

Let us advance a step and note the paradoxical truth of the principle that all penalty for the violation of Law is not only inherent and corrective, but actually kindly. Were we able to sever cause and effect, and abolish all punishment that is seemingly severe, violation would continue until logically followed by destruction. If one be idle, and poverty be the natural penalty, the latter is not an "evil," but a corrective monitor. Were it possible to entirely "abolish" it, idleness would continue indefinitely and find its end in decay. All so-called economic ills have bound up with them the rectifying forces of self-correction. The slower we are to learn their lessons, and the more we count them as "evils," the more severe the discipline that will be self-enforced. Antagonism is the most negative of all the negations of the universal Law of Attraction, and it therefore brings a bitter yet still remedial penalty. Whether in individual or combination, capital or labor, rich or poor, high or low, it can only be transmuted into harmony and benefit through the purifying fires which make up inherent and severe retribution.

In the great economy of Law, intelligent and truly altruistic effort will not have for its object the abolishment of penalty, but of that which brings it. It is a legal part of Law that pain follows in the train of violation. Penalty is the shock we feel when we come in collision with Law. It kindly goads us until we come into conformity, but not by a hair's-breadth beyond that attainment. In economics, as elsewhere, it is not the punishment, but the sin which bears it as a fruitage, that needs to be eliminated.

The fact that Law, in its immutable lines, can never be bent nor diverted in the least degree, is all that prevents the cosmos from becoming chaos. Retributory action in every department, physical, mental, or moral, is universally inherent and corrective and never arbitrary or from the outside. Our mistaken and antagonistic attitude towards it reflects our own hostility back. If we were to receive it as a necessary educator instead of an angry opponent, its face, to us, would be transformed so as to express its natural friendliness.

Economic transgressions always bring their remedies with

them, and if the latter were not bitter they would not cure. We often regard financial panics as unmitigated calamities, but they really cleanse the system of the body-politic. They seem severe because they involve not only the economically guilty, but also thousands of the innocent and of non-participants. Human relationship is so intimate and unitary that those who are lawful suffer with and for those who are lawless. The latter also find some degree of succor in the virtues of the former. Such a commingling — though having a superficial appearance of injustice — breaks the boundary walls of the smaller or personal interest; and by its educational revelations brings to view the larger and truer Unit. It shows that there is no selfish element in Law, and that, paradoxical as it may at first appear, it is best that the innocent should divide the penalties of the guilty with them. Law has made no mistake, even though our selfish concern makes her seem unreasonable. If each suffered solely for his own economic transgressions, it might teach him prudence on his own account, but now he finds that his interest is woven into the very warp and woof of society. It is simply impossible for him to live unto himself, even from the standpoint and purpose of self-interest. To help himself through the promotion of the general good, at length reveals the larger solidarity. The fact that transgressors and non-transgressors are inextricably commingled at first seems unfair, and even unjust; but under the light of a truer interpretation it is found to be not only wise but positively beneficent. It is a standing object-lesson of racial unity. Law is incapable of true interpretation under any fragmentary restriction. One's superficial or apparent advantage may suffer from things beyond his own control; but when the innumerable lines of relationship and compensation are traced out, it is finally found that he receives his supremest good encompassed within that of his fellows.

Sometimes at the close of a muggy summer's day, when

the atmosphere is heavy and murky, a thunder-storm comes as a purifier, and the air is made balmy and the face of Nature becomes bright and beautiful. The process is severe, and occasionally involves local distress; so all active advancement has some accompaniment of evolutionary growing-pains. Expansion, extravagance, and speculation would go on until general financial inebriety ensued, were it not for occasional economic cyclones which tear away the false masks that have been thrust upon natural and fundamental principles. When financial depression comes, there is a general readjustment of economic compasses, and the bearings are again correctly taken. The weak spots of a sophistical political economy are uncovered. It would be impossible for Law, as a schoolmaster, to educate men by the use of any milder means.

Even the distress and poverty which everywhere follow in the wake of a monetary crisis have their compensations. They not only stimulate industry and production, but inspire a fraternal spirit and awaken a general altruism. Such a condition warms the chill atmosphere of selfishness, and brings into high relief the claims of man upon his brother man. The sweetness of charity is realized, and society is brought to see and feel that it is an Organism. Man discovers that he cannot live to himself alone, and that, in a vital sense, he is his "brother's keeper." Not only panics, but all economic ills, are monitors that rise up to teach us lessons that we refuse to learn in any easier way.

Rich outward environment does not bring harmony and contentment, even though the world believe the reverse, as indicated by the mad race for power, wealth, and position. Material attainment, however wonderful, will never usher in the Golden Age. The wealth of invention, which has so greatly augmented man's physical accomplishment during the past fifty years, has conferred no additional happiness. Material progress will be utterly barren in the proportion that it becomes an end instead of an accessory. The greatly broadened scale of material comforts only increases man's sullen discontent with his lot. Humanitarians who confine their efforts to the amelioration of physical conditions alone only touch the surface of human misery. Without a higher evolution of character, if every one were housed in a palace, dissatisfaction, rivalry, and restlessness would still be the rule.

Law seems stern and even baneful, when, through our ignorance, we foolishly antagonize it. But we may render it not only harmless, but transmute it into an infinitely powerful ally. He who utilizes steam or electricity in accordance with their own laws, multiplies his physical accomplishment a thousand-fold; but if he disregard their orderly methods, and strive to impose his own notional theories upon them, he will receive the judgment of penalty. The thorough comprehension of Law is therefore the supreme human attainment.

In the popular mind the idea of Law in political economy is largely limited to the law of human legislation. The collective will of society, as expressed in statute books, is by no means identical in its mandates with Law, although government in modern times is increasingly seeking to mould its expressions after the natural order. Man was very slow in recognizing the dominance of Law in the phenomena of matter, and it will take him still longer to comprehend that it is applicable to himself. In past ages, ethical and political economy aimed at social standards framed according to some notional abstract of what human nature ought to be, rather than basing it upon man's constitution as it is. This was owing to the entire lack of any comprehension of the great modern interpreter — Evolution. Ancient philosophers tried abstractly to build out of their own logic that knowledge which is only attainable by a careful observance of the working of laws through existing facts which express them. Plato's Republic, as an ideal State, was outlined from his opinions of what society ought to be. Taking as his starting point the supreme importance of the State, other relations, like family life and affection, were obliterated or left out. Any processes of reasoning which leave Nature and her methods out of the account, or even make her secondary, are of no avail.

The most important mistakes of the world have been its attempts — often well meant — to override or disregard the Established Order. Men think that they can formulate some plan superior and more expeditious than evolutionary processes. Instead of divining that the very lack of uniformity among mankind is a natural and powerful force working toward universal progress, they make comprehensive plans for union through likeness. Law makes up unity of diversity. Organism of every grade is always a harmonious blending of unlike functions.

Nature has no short-cuts, magic, or spontaneity. But her activities, though immutable, are always elastic and comprise the only perfect means to ends. Mind being as amenable to Law as matter, political economy is an exact science. But it becomes so to our consciousness only to the degree that we move parallel to its lines, and trace out the ties which bind events to their antecedent causation. Failing in this we realize discord and confusion.

But Law, though unchangeable in itself, is, to us, progressive as we make advances onto its higher planes. Our progression comes from the growing supremacy of the higher over the lower motives in the human mind, and the former are no less natural than the latter. The Ideal towards which our faces are turned is as positive as the Real which is at our side. The difference is only in evolutionary location, the former being farther along the great highway.

Human legislation has largely been an effort to impose collective notional Will — at the time ruling — without much search for the true Criterion. Just where it has deflected from this standard it has failed.

But there is a notable dissimilarity between natural and legislative law which is significant. Human enactment is almost entirely negative. It consists of a comprehensive and ever repeated, "thou shalt not." Natural Law is positive and, so far as it is observed, negations are left behind. Negation is the absence of any *thing*. Stealing, cheating, and all economic abuses, are lacks or non-recognitions of the normal social economy. They are spots where man fails to interpret the laws that are written within his own constitution. Hence they introduce inharmony and subjectively nullify the Established Order. Selfishness is not natural, in the high and normal sense of that term, but it has the appearance of a law because animality or human immaturity is common.

That which men have in themselves they see everywhere objectively reflected. One who is disposed to cheat sees cheating in the atmosphere around him, until he mistakenly concludes that it is a part of the Established Order. But it is entirely in men, and Law knows it not.

Ideal political economy is the pure natural system unmarred by the clouded consciousness of its daily multiform infractions. Idealism is as profitable in economics as in any other great subdivision of Truth. It consists of holding up the true potentiality or fulness of what already *is*. It is not a vapory uncertainty of the future but the present perfect substance. If men everywhere held the Criterion before them instead of comparing themselves among themselves, general progress towards the normal would be rapid. Ideals are always striving to actualize themselves.

The great desideratum of man is not an ever increasing aggregation of facts, but rather a clearer perception of the outlines of the Normal. We must also dismiss the idea that because abnormity is common, it is Law. Adam Smith made it plain that Nature in the human mental economy had made provisions for every man, so long as he observed the principles of justice, to use both his own industry and capital with the utmost freedom of competition. This harmony with Nature he taught would result in the largest measure of individual and national wealth and prosperity. In proportion as human legislation is more restricted than Natural Law, men are deprived of the power to work out ideal results. Its bungling interference not only defeats the end sought, but deranges all the delicate and elastic forces which, if left to their higher working, would hasten normal development. If it deprives a laborer of a part of his hours, or if his union orders him out of a situation, in either case he loses something of his natural freedom. Unrestricted competition, both for labor and capital, is the only full measure of liberty for the individual and the nation. All artificial forces that seek to install themselves under the plea of special advantage to some class or faction, in their practical working are only new forms of tyranny which retard natural evolutionary advancement. Every fraction must find its good in that of the whole. The human phalanx cannot be turned aside into artificial bypaths, but its way is already smooth along its own natural highway.

Political economy is the outward expression of the play of the forces of the mind. It is like a game of chess; the pieces being moved after the real move has been made in the mind of the player. As the powers that are within man are tamed, controlled, and brought into orderly harmony, all external phenomena, whether of labor, capital, land, or money, will exactly correspond, for the reason that they are secondary and expressive. Mind is the worker, and these are its tools.

The spirit of association must broaden its aims and interests. It may be a positive institution, but if its purpose be unfriendly to the general and greater Unit, to that degree it is normally unlawful.

Intellectual logic is inadequate to the delicate interpretation of Natural Law, and of its articulated adjustment to human affairs. The intuitive faculty being keener, and of higher grade, is however able to make its leadings so clear that they may be translated into outward harmonious expression. Intuition alone is able to put its ear to the ground and distinguish between discordant, even though faint jars, and concordant vibrations. Only that delicate insight which lies deeper than a mere intellectual account of phenomena, can cognize the lights and shades of those fine but immutable golden threads which are shot through the entire social fabric.

The supreme and ideal political economy can only be formulated from the standpoint of racial unity. Any study of combinations, competitions, and co-operation, cannot be exhaustive on the basis of a fragmentary society with divided interests. Only a synthetic interpretation is adequate, because analysis and separation invariably show disproportion. Man is One; and just in the measure that that grand fact is installed in human consciousness, are *all* the natural principles found to be altruistic. Any philosophy of Humanity is incomplete which does not regard it as an *Organism*. Its members, though unlike, have one interest and one order. Any suffering or rejoicing cannot be localized, for its vibrations thrill to the utmost limits.

INDEX.

ABILITIES of men not equal, 162.
Ability, executive, at a premium, 177.
Abstinence necessary to the would-be capitalist, 56.
Accumulation, passion of, a curse, 183.
Action and Reaction, 196-207.
Agitators, professional, 90.
Alms-giving not the best help, 141.
Altruism demands disciplinary penalties, 93.
—— distress and poverty teach, 290.
Antagonism and remedial penalty, 288.
—— between capital and labor, 99.
—— in society assumed to be necessary, 47.
Apprenticeship system, 273.
Arbitration, governmental, 116-120.
—— voluntary, 119.
Association, principle of, misapplications of, 76.
—— spirit of, must be broadened, 294.
Astor estate, 177.

BANKING system, evolution of the U. S., 215.
—— system, national, an outgrowth of need of government, 202.
Banks, national, as monoplies, 216.
—— state, 215.
Barbarism, insecurity of private property a cause of, 153.
Barbarous tribes, currency used by, 218.
Baring Brothers, failure of, and panic of 1890, 204.
Barter unsuited to civilized communities, 212.
Beneficence must be voluntary, 172.

Benevolence not a function of business corporations, 243.
Bimetallism, 218.
—— by natural law, 224.
Black-listing, 106.
"Booms" and panics, 196-207.
Boycott an injury to society, 87.
—— extreme application of, in Australia, 90.
Brotherhood of locomotive engineers, 81.
Business, dull, caused by dishonesty, 247.
—— methods vs. those of sentimentalists, 78.

CANAL, the first English, 258.
Capital and Labor; can they be harmonized? 158-168.
—— benefits all classes, 164, 181.
—— combinations of, 60-72.
—— defined, 159.
—— effect of annihilation of, 78.
—— labor unions antagonistic to, 77.
—— not limited in quantity, 159.
—— short average life of, 180.
See Labor troubles.
Capitalist, difference between corporation and, 240.
—— would-be, must practise abstinence, 56.
Centralization a condition of wide distribution, 244.
—— law of, 186-194.
Character and competition, 35.
—— and extinction of price, 32.
—— and industrial education, 279.
—— defective, of managers the bane of corporations, 247.
—— defective, the cause of failure, 43.

Character, defective, the cause of ills of society, 94.
—— degraded by voluntary dependence, 139.
—— inefficient service deteriorates, 92.
—— labor unions unfavorable to manly, 80.
Charity, mistaken, 141.
—— money given in, increasing, 171.
—— should be scientific, 138.
  See Beneficence and Philanthropy.
Cities, growth of, 189.
Citizenship, good, and moral education, 275.
Civilization and centralization, 188.
—— depends upon property rights, 153.
Classes, interests of, mutual, 47.
Clearing-house certificates, 207.
Coinage, money and, 210-224.
Colleges, deficiencies of work of, 278.
Combinations amongst employers, 106.
—— and natural tendencies, 29.
—— menace of, to government overrated, 193.
—— of capital, 60-72.
  See Monopolies.
Commerce and the American boy, 276.
—— evolution of the large retail establishment, 190.
—— increased by credit, 214.
—— Stewart estate, 178.
Communism, voluntary local, 152.
Competition and Co-operation supplementary, 43.
—— natural monopolies, 68.
—— between parts having like function, 101.
—— blamed for evils, 93.
—— has lowered rate of interest, 179.
—— law of, 17, 34-40.

Competition must govern until an ideal millenium is realized, 154.
—— price and, 27.
—— should be unrestricted, 131.
—— unrestricted, and liberty, 294.
Competitive system a scapegoat, 151.
—— beneficent, 43.
Conciliation superior to arbitration, 119.
Confidence a subjective factor of money, 214.
—— conditions for, 112.
—— lack of, and panics, 203.
—— the life-blood of prosperity, 92.
—— want of, in railroad management, 248.
Consolidation of railroads, 260.
—— cheapens service, 267.
Consumption lowered by high tariff, 229.
Contract, freedom of, the cornerstone of true government, 117.
—— shall freedom of, be destroyed? 91.
Co-operation an element in profit sharing, 104.
—— a remedy for labor troubles, 165.
—— competition an element in, 36.
—— effect of, without competition, 38.
—— law of, 42-48.
  See Association and Profit-sharing.
Corners, 61-72.
—— "French Copper Syndicate," 64.
—— usually unsuccessful, 28.
Corporate management, abuses of, 246-253.
Corporation, the modern, 238-244.
Counterfeits under State banking system, 216.
Credit, 214.
—— high, valuable to States, 222.
Crime and industrial education, 279.
Currency, effect of war of 1861 on, 201.

INDEX.

Currency, relation of volume and purchasing power, 202.
— the cause of panic of 1893, 204. See money.
DEBT and panics, 198.
Debtors injured in the end by cheapened currency, 222.
Degradation caused by idleness, 53.
Demand an element of value, 40.
— supply and, 24-32.
Democratic party, position of, upon the tariff, 229.
Demonitization of silver, 220.
Dependence and poverty, 134-141.
Despotism in the guise of philanthropy, 117.
Discontent, cause of amongst manual laborers, 162.
Discrimination in rates by railroads, 264.
Domestic service and competition, 37.
— false pride about, 93.
Drudgery eliminated by education, 278.
Duties, incidents of, 235.
— kinds of, 234.

ECONOMIC evils, cause of, 15.
See Ills of society.
Economy. See Social economy.
Education, conventional, and ethical training, 274.
— inadequate, 273.
— of individual the basis of true co-operation, 46.
— the great need of the wageworker, 110.
See Industrial education.
Eight-hour system in Chicago, 83.
Employees, advantages of having, shareholders, 104.
— how to secure allegiance of, 102.
— not machines, 100.
— obligations and privileges of, 108-113.
Employer, aim of laborer to become an, 55.

Employers and profit-sharing, 98-106.
Emulation, influence of, 38.
Enterprise the great stimulus to, 154.
Environment as a factor of happiness, 290.
— effect of, upon American workman, 231.
— good of individual as related to, 287.
Ethical training necessary, 274.
Evil defined, 286.
Evils. See Economic evils.
Evolution, mind as a factor in social, 57.
— of the large retail establishment, 190.
— the great interpreter, 291.
Excellence, influence of labor unions against, 79.
Exchanges, intrinsic value of money alone considered in foreign, 214.

FACTORY legislation, 129.
Failure due to defects in character, 43.
Faithfulness stimulated by profitsharing, 102.
Farmer and the tariff, 233.
Fiat element in ideal currency, 212.
Fortunes, can, be limited? 172.
— great, and development of railroad system, 175.
— more difficult to acquire than formerly, 178.
— unearned, 248.
Freedom of contract, shall, be destroyed? 91.
— subverted by labor unions, 81.
Free list, 235.
Freight rates, 263.
— a century ago, 259.
— decline in, since 1865, 265.
— effect of reduction in, 69.
—, normal, defined, 266.
See Railroad tariffs.
"French Copper Syndicate," 64.
Futures, speculation in, 69.

GOOD defined, 286.
—— of individual as related to environment, 287.
Gold, qualities commending, for use in coins, 216.
—— ratio of value of, to silver, 219.
Government, menace of combination to, overrated, 193.
—— our free, favorable to production, 58.
—— ownership of railroads, 250, 269.
" Granger laws," 126.
Greeley, Horace, upon voluntary local communism, 152.
Growth, law of, 14.

HAPPINESS, environment as a factor of, 290.
—— not a necessary effect of wealth, 161.
—— proportioned to merit, 93.
Harmony between capital and labor, can, be realized? 158-168.
—— conditions of social, 46.
—— how to secure, 100.
—— intuition and social, 295.
See Conciliation.
Health and overwork, 52.
Heredity, laws of, tend to disperse wealth, 180.
Hours of labor, 92.
—— in Chicago, 83.
—— statutory regulation of, 128.

IDEALISM, natural law and, 284-295.
Ideals and progress, 293.
Idleness a torture, 111.
—— bad effects of, 53.
Ignorance a cause of want, 94.
Ills of society, legislation involved to cure, 147.
—— no panacea for, 164.
—— the outcome of defective character, 94.
See Economic evils.
Improvidence caused by idleness, 53.

Independence defined, 54.
Industrial education, 272-281.
—— an agency for cure of social ills, 94.
" Industrials," speculation in, and panics of '90 and '93, 204.
Industry, influence of labor unions against, 79.
—— purpose of legislation and political science to protect, 154.
Inflation a calamity to labor, 174.
—— effect of, on wages, 222.
—— eras of, make accumulation of wealth easy, 173.
Inheritance, laws of, tend to disperse wealth, 179.
Intemperance caused by idleness, 53.
Interdependence of mankind, 289.
Interest has declined with increase of wages, 161.
—— declining, 179.
—— effect of legislation upon, 27.
—— rate of, paid by railroad investments, 268.
Interstate commerce law, 262, 265.
Intuition and social harmony, 295.
Investments, effect of unrest upon, 91.
—— made difficult by dishonest corporate management, 248.
Investors and railroads, 260.

JERVIS, JOHN B., upon railroad property, 260.

KNOWLEDGE, useful, defined, 16.

LABOR, all kinds of, interdependent, 111.
—— and capital, can, be harmonized? 158-168.
—— and production, 50-58.
—— causes of over-supply of, 92.
—— combinations of, 74-95.
—— defined, 159.
—— demand for, increases with advance of civilization, 161.
—— does, receive a fair share of the product? 162.
—— indispensable to manhood, 93.

Labor, inflation a calamity to, 174.
—— manual, not degrading, 53.
—— money stored up, 212.
See Hours of labor.
Laborers and their champions, 6.
—— cause of discontent amongst manual, 162.
—— labor unions tyrannize over unorganized, 87.
—— standard of living of, in the U. S., 231.
—— unorganized, ignored, 88.
See Employee.
Laboring man, competition and, 39.
—— meaning of term improperly restricted, 39.
Labor troubles, profit-sharing the remedy for, 104, 166.
—— the spirit that harmonizes, 101.
See Harmony.
—— union, the ideal, 94.
—— unions and American youth, 274.
—— bad effects of, upon laborers, 55.
—— effect of, upon wages, 30.
—— fight of, for exclusive recognition in Australia, 89.
—— put a premium upon incompetency, 39.
—— when, are mere animal co-operation, 45.
Labor-value, how, is fixed, 160.
—— the sole basis for protection, 231.
Land, effect of governmental ownership of, 154.
—— made valuable by railroads, 176.
Law defined, 286.
—— foundation of human, 19.
—— place of, in human life, 291.
—— universality of 7,285.
See Natural law.
Laziness a cause of want, 94.
Leclair's experience with profit-sharing, 102.
Legal tender in colonial times, 212.

Legislation and the panic of 1893, 206.
—— an element in ideal currency, 212.
—— economic, and its proper limits, 122–131.
—— generally negative, 293.
—— in favor of debtors harmful, 222.
—— invoked to cure social ills, 147.
—— lax in dealing with abuses of corporate management, 248.
—— limited power of, 14.
—— not identical with natural law, 291.
—— purpose of, to protect industry, 154.
—— reform of abuses in railroad management by, 250.
—— restrictive railroad, detrimental, 265.
—— special, dangerous, 117.
—— tendency towards general, affecting corporations, 242.
Liberty requires unrestricted competition, 294.
Living, standard of, of laborers in the U. S., 231.
Lock-outs, 106.
"Long and short haul clause," 265.

MACHINERY, effect of labor-saving, 56.
—— on tariff, 231.
Man, the ideal, 53.
Management, abuses of corporate, 246–253.
—— disadvantages of governmental, 124.
Manual employment and false pride, 274.
—— the American boy, 276.
—— Training. See Industrial education.
Margins, 70.
Mental force exceeds manual force in its reward, 162.
Mind an increasing factor in social evolution, 57.
—— increases production, 183.

Misery, pictures of, do not arouse men to action, 136.
Money and coinage, 210-224
— stored-up labor, 212.
See Currency.
Monometallism, silver, 221.
Monopolies, causes and tendencies of, 188.
— national banks as, 216.
— natural, and competition, 68.
— rates of railroad, governed by natural law, 267.
See Combinations, Trusts, Standard Oil Co., and Western Union Telegraph Co.
Municipalism, limits of, 123.

NATIONALISTS, plan of the, 147.
Nationalization of railroads, 250.
— chimerical, 269.
Natural law and idealism, 284-295.
— characteristic of, 13.
— reliability of, 16, 20.
— scope of, 14.
— supremacy the result of knowledge of, 57.

ORGANISM and unlike functions, 292.
— society an, 295.
Organization, more efficient, the basis of corporations, 240.
Over-production, how to abolish, 112.
— influence of tariffs on, 233.
Overtrading a cause of panics, 198.
Overwork harmful, 52.

PANACEAS not a remedy for the ills of society, 15.
Panic and distrust of fiat element in money, 214.
— "Booms" and, 196-207.
— caused by labor unions, 91.
— of 1893, 215.
Paternal government cannot harmonize capital and labor, 165.
— fosters dependence, 140.
Paternalism and competition, 39.

Pauperism and poverty differ, 140.
Penalties beneficent, 287.
Philanthropy may make despotism, 117.
— on business principles; the Pullman Co., 105.
— true, requires the study of causes, 94.
See Charity.
Political economy and exact science, 292.
— and the place of mind, 294.
— conventional, 6.
— defined, 18.
— factors in, 204.
— relation of, to the mind, 286.
— standpoint of ideal, 295.
— study of, 16.
Political science, purpose of, to protect industry, 154.
Population flows to cities, 187.
Postage, effect of lower rates of, 192.
— rates of, 258.
Poverty and competition, 37.
— are the poor growing poorer and the rich richer? 173.
— dependence and, 134-141.
— no panacea for, 164.
— not an evil, 288.
Precious metals, parity of, 223.
Price, competition and, 27.
— effect of trusts on, 40.
— equalizes supply and demand, 26.
— extinction of, 32.
— fluctuations minimized by modern conditions, 179.
— imposition of artificial, immoral, 62.
— of bread regulated by law, 127.
— of railway service, 263.
Prices, artificial, short-lived, 28.
— fall in, 203.
Product, does labor receive a fair share of, 162.
— effect of prevailing fallacies on, 39.

Product of manual labor, increase in, 151.
Production a minimum under socialism, 149.
—— distribution of instruments of, 231.
—— increases with brain power, 183.
—— in England and the U. S. compared, 232.
—— labor and, 50-58.
—— natural unit in, 99.
—— not solely the result of physical labor, 109.
—— our free government favorable to, 58.
—— the great stimulus to, 154.
Professions overcrowded, 273.
Profit-sharing, employers and, 98-106.
—— the remedy for labor troubles, 166.
Progress accelerated by competition, 38.
—— and ideals, 293.
—— conditions of, 52, 54.
—— depends upon diversity, 292.
—— key to, 19.
—— law of ecomonic, 58.
—— socialism fatal to, 146.
Progression, how, takes place, 292.
Property rights the basis of civilization, 153.
—— rises in value during periods of inflation, 174.
Prosperity depends upon confidence, 92.
—— jeopardized by strikes, 82.
—— permanent, 223.
—— precedes panics, 200.
—— results from obedience to law, 47.
—— sacrificed by frequent changes in tariff, 230.
—— the great stimulus to, 154.
Protection, tariffs and, 226-236.
Public opinion lax in dealing with abuses of corporate management, 248.

Pullman Car Co., successful experiment of, in benefiting workmen, 105.
Purchasing power of money, increasing, 217.

QUALITY an element of value, 40.

RAILROAD building, abnormal, the cause of panic of 1873, 202.
—— evolution of the, 256-270.
—— management often dishonest, 249.
—— system, development of, and great fortunes, 175.
—— tariffs and legislation, 28.
—— tariffs by legislation, effect of fixing, 126.
—— tariffs fixed by natural law, 127.
See Freight rates.
Railroads, governmental management of, 125.
—— governmental ownership of, 250, 269.
—— strike of 1886 on Gould system of, 82.
—— wrecking and reorganizing, 251.
Raw material? what is, 234.
Reaction, action and, 196-207.
Real estate, is, better than other investments? 177.
Realism will not cure the ills of society, 136.
Reciprocity and exports, 233.
Republican party, position, upon the tariff, 229.
Retail establishment, evolution of the large, 190.
Retaliation provoked by unfriendly tariffs, 233.
Retribution, 288.
—— inevitable, 20.
Roads of the Roman Empire and of Europe in the Middle Ages, 257.

SCHOOLS, criticism of common, 274.
Self-interest and social economy, 13.
Selfishness, law of, 287.

Sentimentalism leads astray, 15.
—— methods of, *vs.* those of real business, 78.
—— opposed to natural law, 152.
Service, effect of legislation on rates of, 28.
Shareholders, how interests of, can be protected, 252.
—— relation of, to corporations, 242.
—— restrictive railroad legislation unjust to, 265.
—— victims of a false system, 249.
"Sherman law," 205.
Short sellers, 70.
Shutting down involves loss, 193.
Silver, demonitization of, 220.
—— qualities commending use of, for coins, 216.
—— question, 221.
—— ratio of value of, to gold, 219.
Slum population, character of, 135.
Slums need optimism, 137.
Social economy, motive of, 13.
—— evils. *See* Ills of society.
Socialism as a political system, 144-155.
—— basic fallacy of, 181.
—— genius of, 154.
—— influence of labor unions towards coercive, 88.
Socialistic experiments, voluntary local, 151.
—— party, animus of, destructive, 148.
Social system, relation of, to development, 14.
—— War. *See* Antagonism.
Society an organism, 205.
Solidarity of the human race, 289.
Speculation and panics, 198.
—— stimulated by dishonest corporate management, 248.
Spoils system opposed to governmental management, 125.
Standard of living of workmen in the U. S., 231.
—— Oil Co., 191.

Standard Oil Co., a case of successful competition, 37.
—— why the, is successful, 67.
State interference, 122-131.
Stewart estate, 178.
Stockholder. *See* Shareholder.
Stock-watering, 268.
—— does not affect the rates of service, 192.
Strikes, consequences of, illustrated, 81.
Success comes by natural law, 110.
—— not due to chance or luck, 180.
—— secret of, 112.
Suffering, causes must be studied to remove, 94.
Sugar trust, 66.
Supply and demand, 24-32.
—— apply to charity, 138.
—— govern purchasing power of currency, 202.

Tariffs and protection, 226-236.
Technical education. *See* Industrial education.
Telegraph rates, effect of reduction in, 69.
Trade schools. *See* Industrial education.
Transgressions, economic, bring their remedies, 288.
Transportation, 265-270.
—— rates, pool, short-lived, 28.
See Railroads.
Travelling facilities by rail, changes in, 261.
Trusts, 61-72.
—— and prices, 40.
See Monopolies.
Truth always beneficent, 17.
—— conditions for finding, 17.
Tulip mania in Holland, 199.

Under-valuation and *ad valorem* duties, 234.
Unearned increment, is there an, 177.
Unionism. *See* Labor unions.

Union should be between parts forming productive unit, 102.
Unions, unnatural, 99.
Unselfishness, how, is unfolded, 14.
Usury laws injurious, 128.

VALUE, effect of fluctuations of, 174.
—— how, is affected, 40.
—— how, is conferred, 72.
—— ideal currency must have intrinsic and stable, 212.
—— sentiment a factor in, 204.
Values, natural law weighs, 110.
Virtues, competition applies to, 36.
—— should be taught, 275.

WAGES and labor unions, 40.
—— education the basis of increased, 110.
—— effect of labor unions on, 30.
—— when, are increased by pressure of labor unions, 84.

Wages, how, are increased, 31.
—— increase of, accompanied by decline of interest, 161.
—— rate of, advancing, 86.
—— regulated by supply and demand, 80.
—— should not be fixed by the state, 118.
Want, causes of, 94.
Watering stocks. *See* Stock-watering.
Wealth, a curse when dishonestly acquired, 112.
—— and its unequal distribution. 170–183.
—— great accumulations of, benefit the poor, 160.
—— happiness not a necessary effect of, 161.
Western Union Telegraph Co., 68.
—— monopoly tends to cheapen service, 191.
Work a blessing, 110.
—— theory of minimum, 92.

SECOND EDITION

IDEAL SUGGESTION
THROUGH
MENTAL PHOTOGRAPHY

A Restorative System for Home and Private Use Preceded by a Study of the Laws of Mental Healing

By HENRY WOOD

AUTHOR OF "GOD'S IMAGE IN MAN" "EDWARD BURTON" "NATURAL LAW IN THE BUSINESS WORLD" ETC.

Cloth $1.25

Part I. of this work is a study of the *laws* of Mental Healing, and Part II. embodies them in a restorative system, formulated and arranged for home and private use. Visionary and impracticable aspects of the subject are eliminated, and a scientific basis is found. The book is not technical, but thoroughly plain and concise, and will prove a boon to invalids and a valuable addition to the substantial literature of the subject.

PRESS OPINIONS

B. O. FLOWER in " The Arena "

"Recently, however, some scholarly and finished works have appeared, which will take high rank as literature, and will doubtless hold a permanent place among the thoughtful and thought-inspiring books of the present generation. Notable among them are Professor Wait's 'Law of Laws' and Henry Wood's latest work, 'IDEAL SUGGESTION THROUGH MENTAL PHOTOGRAPHY.' This last work is one of the most charming volumes of essays of recent years. Henry Wood is the Emerson of the new metaphysical thought, and in his writings there is a certain wealth of thought and felicity of expression not found in Emerson. I know of no American essayist to-day who clothes his ideas in such a wealth of rhetorical expression, and who is never verbose, as Mr. Wood. If his style is florid and poetic, there is never any superfluous word or sentiment introduced for artistic effect. He has a magnificent command of language, and expresses his ideas with rare felicity, which makes anything coming from his pen delightful reading, even though one may differ radically with the thought expressed. In the present volume Mr. Wood appears at his best as an essayist. Indeed, in many chapters he seems to even surpass any former work.

"'Ideal Suggestion' is divided into three parts. The first treats of the Laws of Mental Healing. In this section are five chapters, which for clearness, conciseness, fluency of expression, have rarely been equalled. The subjects discussed are: The Obstacles to Progress, The Body, The Power of Thought, Planes of Consciousness, Inferences and Conclusions.

"Part second deals with Ideal Suggestion, and contains practical suggestions for those who wish to treat themselves along the line of mental healing. The third division contains Meditation and Suggestion, in which are given twenty short lessons, and an equal number of thought phrases to be held mentally.

"The volume is handsomely printed in large type on heavy paper, and beautifully bound. It is a work which every person interested in metaphysical healing should possess, and will be an admirable volume to loan persons interested in this thought."

THE BALTIMORE METHODIST expresses its opinion thus freely

"Henry Wood, author of 'God's Image in Man' and other interesting books which have gone through a number of editions, has just given to the public, through the well known Boston publishers, Lee and Shepard, another volume, entitled *Ideal Suggestion through Mental Photography*. Anything Mr. Wood writes will find numerous readers because of his style, if for no other reason. We doubt if he is excelled in this particular by any other English writer. He is, however, more than a master of style. He is a profound reasoner, plunging fearlessly into the depths of mental philosophy and psychology. As a study of the laws of mental healing, his book is a good one, but the attempt which he makes to formulate from these laws a restorative system by which individuals may eradicate disease without the use of any other means than mental causation we think rather visionary. That disease has its seat in the mind alone, and not in the body, we are not prepared to credit. Other animals than man suffer from disease. Will 'ideal suggestion' meet their case? We believe that both mental condition and faith will contribute much to recovery from disease, but we do not believe that either will avail anything where material means are requisite and available. We suspect this volume, like others, by the same author, is but an outgrowth of 'the new theology of evolution.' It is good reading, but poor gospel."

BOSTON TIMES

"Mr. Wood is sufficiently well known as an 'unprofessional,' conservative psychologist to ensure thoughtful and respectful treatment of any effort he may make in this field; and we do not question that many persons will be richly rewarded for the attention they give to this, his latest work. The ordinary text-book on mental healing is an ill-digested medley of occultism, metaphysics and jargon. A special illumination is required to read it, and not even the author understands it. But Mr. Wood has taken pains to be lucid, is eloquent at times, and is always direct, fair-minded and hopeful. The most confirmed materialist might read with a degree of pleasure, and since investigation and belief seem to be tending in the direction of Mr. Wood's theories, it is likely that his comprehensive precept will largely lead to practice."

PROVIDENCE JOURNAL

"Lest the reader should not easily discover from his title, the subject of Mr. Wood's volume, one may explain it, roughly, as mind cure — an attempt to find in mental causation both a scientific and a spiritual basis for bodily healing and health. The book is not too technical, it is interesting and suggestive even to those who believe that the great commission, 'Preach the Gospel and heal the sick,' has existed ever undivided in the church; who do not believe that a direct answer to prayer implies that God is subject either to change or improvement, or that a belief in the sacredness of relics, and the possibility of miracles wrought through them is what he terms 'pure superstition.' The first portion of his book is devoted to a study of the laws of Mental Healing; the second to restorative system of meditations arranged for private use. Mr. Wood's interest in the subject is unprofessional, and he is well known as an interesting, thoughtful writer upon cognate topics."

CHICAGO CHRISTIAN METAPHYSICIAN

"This volume will wear well, it has staying qualities, the plan is readily understood and can be used by a novice or a profound Scientist, the benefit realized differing only in degree since its use will always result in some good spiritually, mentally and physically. We heartily commend this latest, best and most helpful book by this thoughtful, progressive author. This is a book of purity and health, scientific and practical.

BOSTON IDEAS

"Mr. Wood is of a nature that enables him and his writings to materially assist in the promulgation of the practical enlightenment which the truths of mental-healing (so long judged by the errors of its representatives) alone can give. An absolute truth eventually expresses itself; it is inevitable. And those whose aroused spirits have perceived and absorbed to ever so small a degree that spark which vitalizes human consciousness will necessarily keep on seeking new light from the same source; and those inclined to be somewhat 'irrational' in their flights will gain immeasurable good from Mr. Wood's clear, incontrovertible statements.

"The value of clear, direct announcement, insistently reiterated and kept before the thought is embodied in the system of 'meditations' which occupies the latter portion of the book. It will strengthen a negatively inclined mind and invigorate or leaven a positive one."

PORTLAND TRANSCRIPT

"That the mental attitude affects the physical condition of humanity is an indisputable fact. That as a result there has been a great deal of 'mind-cure' knavery is also a fact. But there are underlying scientific principles in regard to this matter that are now receiving attention. We are glad to call the attention of our readers to a work upon this important subject by a man who is well known as a careful and capable writer upon psychological and metaphysical topics. The title of the book is *Ideal Suggestion through Mental Photography*. The author is Henry Wood, whose other books are 'God's Image in Man,' 'Natural Law in the Business World,' and 'Edward Burton,' a novel. He is an independent investigator and has given the subject much study besides having had unusual personal experience. The book should prove of great value to invalids if they are only intelligent enough to profit by its wise advice and are not those unfortunates who are 'enjoying poor health,' as the saying is."

Cloth Price $1.25

Sold by all booksellers and sent by mail on receipt of price. Catalogues sent free

LEE AND SHEPARD Publishers BOSTON

AN INTERESTING BOOK.

GOD'S IMAGE IN MAN.
Some Intuitive Perceptions of Truth.

BY HENRY WOOD,
BOSTON, U.S.A.,
AUTHOR OF "IDEAL SUGGESTION," "EDWARD BURTON," "THE POLITICAL ECONOMY OF NATURAL LAW," ETC., ETC.

In Cloth, 258 pages, $1.00.
Sold by all Booksellers, or sent, postpaid, by the Publishers, LEE AND SHEPARD, Boston, on receipt of the price.

CONTENTS.

I. *The Nature of God.*
II. *Revelation through Nature.*
III. *Direct Revelation.*
IV. *Biblical Revelation.*
V. *Revelation through the Son.*
VI. *The Universality of Law.*
VII. *The Solidarity of the Race.*
VIII. *Man's Dual Nature.*
IX. *The Unseen Realm.*
X. *Evolution as a Key.*
XI. *From the Old to the New.*

"Its pure and elevated style is wonderfully attractive. This volume is one of rare value." — *Boston Traveller.*

"A notable treatise on the new theology of evolution." — *Brooklyn Eagle.*

"It is certainly instinct with spiritual vitality. It is filled with the light which the scientific method has kindled." — *Boston Home Journal.*

"An honest, able, and promising effort to free faith from unnecessary incumbrances." — *New York Independent.*

"Mr. Wood has done us a service, and we trust that many will receive from the same and subsequent volumes spiritual quickening." — *The Critic* (New York).

"A volume full of deep and suggestive ideas from the standpoint of the theology of the divine immanence." — *The Christian Union* (New York).

"The book cannot fail to prove helpful in the renaissance of Christianity that is going on in our day." — *The Unitarian* (Boston).

"The book is profoundly religious in tone, and breathes the spirit of the so-called new orthodoxy." — *The Review of Reviews.*

"The fact that the unseen universe is as accessible from America as from India is one which the Western thinker has been slow to grasp, and Mr. Wood has been perhaps the first to present it frankly yet delicately with an absolute absence of that occult assumption which has done more than anything else to prejudice the intellectual world against the investigation of psychic questions, involving an intimate acquaintance with one's own soul and its possibilities." — *Kansas City Mail.*

"The book is vigorous and suggestive." — *San Francisco Chronicle.*

"Mr. Wood writes for thoughtful men on serious topics." — *Chicago Herald.*

"One need not always agree with Mr. Wood in his theories to take pleasure in reading his books. He is never dull; he is always reverent when speaking of things which others revere, though some of these things may be regarded by him as groundless superstitions; there are scores of excellent thoughts flowing from his pen, which serve to inspire one to better things than the common round of every-day grind. His 'Edward Burton' was an uplifting, religious novel, which has passed through several editions and will pass through many more, for it pleases the always-increasing American-nobility class — the readers whose motto is always and everywhere *noblesse oblige.*" — *British American Citizen* (Boston).

"Mr. Henry Wood, who has gained many readers by his 'Natural Law in the Business World,' and by his articles upon religious subjects which have appeared in the magazines, has justified the hopes of his admirers in his last work, 'God's Image in Man,' in which he discusses some of the most important theological questions of the day in a most common sense manner. The author is an original thinker and depends for his statements upon neither dogma nor prejudice." — *Boston Courier.*

"The religious world could better afford to lose whole volumes of dreary commentaries and reflections among the tombs, and such like aids to future happiness and present somnolence, than one page of such illuminating and inspiring writing as this." — *Charleston News and Courier.*

"It is both a pleasing and profitable book." — *Chicago Inter-ocean.*

"The book glows with both beauty and power." — *Ohio State Journal.*

"Mr. Wood is a keen and logical thinker, and a lucid and forcible writer." — *The Beacon* (Boston).

"This new book, by Henry Wood, is the product of an intuitive perception of Truth. It presents the principle of Divine Science in an entertaining style, by illustrating the problem of Life in various examples, and in a manner that will prove most interesting and instructive to all thinking people." — *Harmony* (San Francisco).

"The author does not follow any strict logical or philosophical method, but gives free rein to the imagination, and his style is poetic rather than dryly argumentative. He is broad, catholic, and progressive in his views of religion, and logical. The volume is, on the whole, an earnest, catholic, thoughtful exposition of modern ideas of religion and man's relation to the universe; and many who have been hampered by the trammels of mediæval thought may find help in this book." — *The Christian Register* (Boston).

Press Notices of another of MR. WOOD'S *Books.*

EDWARD BURTON.

AN IDEALISTIC METAPHYSICAL NOVEL.

"'Edward Burton' would be called a religious novel. The fundamental thought is the outworking of souls toward light and love from the bondage of oppressive dogma and unreasoning belief. But, unlike many religious novels, the story is not dull, nor does the movement drag." — *The Christian Union* (New York).

"A very powerful story, which holds the reader's attention from beginning to end. Into a pretty love-idyl the author has woven a vigorous account of the influence exerted by the numerous systems of theology, ethics, and sociology, which in our day excite so much attention." — *Peterson's Magazine* (Philadelphia).

"There are some admirable character studies, among them being a snobbish 'milord,' a German Anarchist, and a liberal-minded clergyman of keen spiritual insight and refinement of thought and feeling. The ideals are high, and the book is altogether a stimulating and developing piece of work. The author has already made a wide reputation for himself by his book entitled 'Natural Law in the Business World.'" — *Public Opinion*, Washington, D.C.

"It is difficult to find words to fully express the pleasure we are sure will be derived — at least, by those interested in the search for absolute truth — from the novel entitled 'Edward Burton,' written by Henry Wood. It may be termed a religious novel in the sense that it teaches the purest, truest, most unbiased, most truly practical 'religion' of any novel we have known written on the subject. It *cannot* be termed religious in the sense of subscribing to any creed, or of lending authority to any code whose only life lies in the externals of things. The attempt to put into concrete form our impressions of 'Edward Burton' so leads us off into the so infinitely many and mingled thoughts which each and every page of the book calls forth that we feel our most sensible word is, 'Read for yourself' — which we can say with all-heartiness." — *Boston Times*.

"'Edward Burton' is a delightful book in all its sketches of out-door life; the sea, the streams, the woods, the mountains, shady nooks, walks by moonlight, the varying influence of the weather, and all the voices of Nature are brought to us with charming reality and in wholesome, generous plenty. As a story, it leaves a pleasant after-taste in one's mind. The outcome is neither tragic nor disappointing; the best people are not dragged into shreds and left on the harrow's teeth, nor are the villains successful, as things must be, you know, in the high art of realism. We have given considerable space to Mr. Wood's book, because it is an unusual and a thought-provoking work. The influence of a thinker who is a deep delver, and who is always serious and earnest, comes out of the story, or rather from between the links of the story." — *N.Y. Independent*.

"It is refreshing to take up a book that is emphatically a book of ideas." — *The Writer*, Boston.

"The descriptions of natural scenes are fine, and cause us to breathe their very air. Whether of the seashore or the mountain, the tone of the novel is really that of a romance clothed in familiar incidents. The heroine, Helen Bonbright, by indefinable touches, is a nobly idealized but individual type of spiritual womanhood." — *Andover Review*.

"His comprehensiveness of character, and the author's power of imparting the same, show much ability and tact. The sentiment therein expressed will echo and re-echo in many a heart." — *Falmouth Local*.

"The excellence of the language is one of the chief charms of the work, and the careful reader cannot but be impressed with the idea that the author is a close observer, a profound thinker, and an excellent scholar." — *Sacramento Bee*.

"Few books of the character of 'Edward Burton' have been issued that have received more attention from men of thought and students of the idealistic school than this novel is destined to attract." — *Wheeling Intelligencer*.

"It is finely written, and, although the author does not, perhaps, intend to rebuke anything or anybody, it is, nevertheless, a rebuke to much of the realistic, analytic, and pessimistic literature of the day." — *Toledo Blade*.

"The plot is artistically excellent, and its working, as well as the literary style, is easily marked as elegant." — *Nashville American*.

"No one can read it without having his understanding enlightened and his aspiration toward the higher life quickened and increased. The book is written, too, in an easy and graceful style, and contains enough of pleasant and romantic incident to interest the ordinary reader of fiction." — *New Christianity*, Philadelphia.

DREAMS OF THE DEAD

By EDWARD STANTON, with an introduction by EDWARD S. HUNTINGTON
Third edition   Paper 50 cents; cloth $1.00

Dealing with the most abstruse problems that vex, perplex, and fascinate the soul of man, it does not pose as authoritative; it is not declarative as a whole; but, on the contrary, it is replete with a fine feminine quality, the delicate, yet gigantic power of suggestiveness, . . . as this book shows, there may be sermons in dreams as well as in stones. . . . It is rare to find a book pretending so little, yet lending so much in the way of suggestion, as this "Dreams of the Dead."

In calling attention to such a work there is no need of falling into the weakness of cheap superlatives. Friendly and unfriendly critics are agreed that in many ways it is a remarkable book. Yet, in the ordinary sense of the word, it is not a novel, either in art, dimension, or aim. It takes the form of simple narrative, and purports to give an account of certain psychic experiences of the author. The style in which the book is written, perhaps, contributes in some measure to this effect. It is appropriate and beguiling. It has none of the falsetto of the sensationalist; rhapsody and rant are excluded; even the exclamatory is absent. It is pervaded by the sombre gray, the local color, if one might so call it, of the subject. It is sober and orderly, but with an easy, flowing movement — caught, one would think, from the gliding ghosts — that gently lifts the reader off his feet and carries him along whether he wishes it or not.

WHICH WINS

A Story of Social Conditions   By MARY H. FORD   Paper 50 cents; cloth $1.00

Mrs. Ford must be credited with the authorship of one of the most profoundly interesting tales of American life, which not only breaks into fresh fields and pastures new, but traverses some of the old paths with the tread of a master. There are many currents in the narrative, but they all are skilfully made tributary to the main stream. The villanies of the financial oppression with which the farmers of prairie States have been visited; the ills of the bonanza farms; the injurious discriminations of the railways — all are vividly set forth with full knowledge of the subject. It is one of the timeliest, strongest, most noteworthy sociological fictions of our day. — *Detroit Journal.*

PECULIAR — A Hero of the Southern Rebellion

By EPES SARGENT   New Edition   Paper 50 cents; cloth $1.00

A story written at the time of the Civil War, and by one who was so active in all the anti-slavery movements, cannot fail to be of interest to the present generation, although it deals with times and people so different from the present that it may seem improbable. The author treats of Southern life during slavery days and the war; and many lessons of manliness and courage are presented of those to whose patriotism, bravery, and sacrifices were due the preservation of the Union.

The basis upon which the story is founded has been swept away, and it is all the more agreeable on that account. It is because the events of former years pass so quickly into obscurity that it is well for us to read such books as "Peculiar."

Books Upon Various Subjects

THE BLIND MEN AND THE DEVIL

By "PHINEAS" Paper 50 cents ; cloth $1.00

"The Blind Men and the Devil" is in the nature of Bellamy's "Looking Backward," differing in this, that while Bellamy's work describes a better condition of labor, this volume assumes to give, in the form of allegory, the condition of labor as it is. "The Blind Men" of the title are the workmen of the present, and the "Devil" is money. The leading character of the story is a journalist condemned to manual labor with other workmen. His experience is given in the style peculiar to the book, in vivid diction, and with an art at putting things that will interest readers fully as much as "Looking Backward." — *National Labor Tribune.*

LAURENCE GRÖNLUND'S WORKS

THE CO-OPERATIVE COMMONWEALTH

An exposition of Socialism By LAURENCE GRÖNLUND A Revised Edition Paper 50 cents; cloth $1.00

People who are curious to know just what Modern Socialism is — what are its dreams, its repressed desires, its plans and expectations for the future, its passionate folly, its mad hatreds, its exalted enthusiasm — can scarcely do better than to read Laurence Grönlund's "Co-operative Commonwealth." — *Boston Journal.*

Mr. Grönlund has re-written his work, and added such points as have developed, in the theories advanced, since the work was first published.

CA IRA! Danton in the French Revolution

By LAURENCE GRÖNLUND Paper 50 cents; cloth $1.00

This work, which is not so much a biography of Danton as a study of the French Revolution, regarded as a preparation for what is yet to come in the revolution of society, the fifth act in the drama being, in the view of the author, the co-operative commonwealth, to which he looks for the solution of the social questions of the day, will command the attention of the thoughtful.

OUR DESTINY The Influence of Nationalism upon Religion and Morals

By LAURENCE GRÖNLUND Paper 50 cents; cloth $1.00

It is full of sublime truth and meaning, combined in a system evolutionally derived, to a degree remarkably satisfactory to broad, rational thought and perception. Every thinker will find "Our Destiny" interesting in every sense.

It is a powerful appeal, written straight from the heart of a living man of profound sympathy and no small intellectual capacity. It probably foreshadows a coming conflict between that section of Socialists which is animated by religious conception, and that other section which is purely materialistic and ultra-revolutionary. — *London Daily Chronicle.*

Lee and Shepard's List

SPEECHES LECTURES AND LETTERS

By WENDELL PHILLIPS   First series, with portrait   Library edition $2.50   Beacon edition $1.50

SPEECHES LECTURES AND LETTERS

By WENDELL PHILLIPS   Second series, with portrait   Library edition Price $2.50   Beacon edition $1.50

We do not know where to turn to a volume that touches all the great thoughts of humanity at more points, or more deeply, than this collection of the utterances of Mr. Phillips upon the different occasions when he was asked to address his fellow-citizens. They reveal the whole man. They indicate his moral and intellectual position as nothing else could. We are glad to learn that we are to have something more in connection with Mr. Phillips's personal history, and that in a short time there will be added to these two volumes a series of speeches and selections which have not before been published, and which will bring out in a stronger light his relations to the anti-slavery movement and the growth of his views and sentiments and the development of his power and fame as an orator. — *Boston Herald.*

ONOQUA

An Indian Story   By FRANCES C. SPARHAWK   Paper 50 cents; cloth $1.00

It is a powerful but painful story, and the cause it advocates is argued with force and justice. Our treatment of the Indians is cruelly iniquitous, and the author is right when she asks indignantly: "How long are we to hold them back from our opportunities, which every other individual may grasp wherever he can find them? How long are arid acres, which they have no means to irrigate, to be considered the sole requisite of these people for citizenship? In a land full of arts and manufactures, how long is the cordon of reservation, like the Libby death-line, to imprison this race, full of mechanical and artistic skill? Who will free the Indians? Only Indians who are free themselves, as only free white men have freed their race." The novel is of exciting interest, and the narrative is earnest, spirited, and picturesque. — *Evening Gazette* (Boston).

SOCIALISM   From Genesis to Revelation

By Rev. F. M. SPRAGUE   Cloth $1.75

In "Socialism," by Rev. F. M. Sprague, is presented a calm and thorough investigation of the question both from the economic and ethical point of view. The author treats the doctrine as an evolution, of which the world of thinkers, and especially of capitalists, *must take heed.* He does not advocate violence or confiscation, but a new arrangement of industrial forces. He shows that as democracy succeeded feudalism and monarchy, so socialism is destined to succeed *capitalism*, and that the inequality and injustice which now prevail must give way to the principles of brotherhood as set forth by our Lord Jesus Christ.

Sold by all Booksellers, or sent, postpaid, by the Publishers, LEE AND SHEPARD, Boston, on receipt of the price.

www.ingramcontent.com/pod-product-compliance
Lightning Source LLC
Chambersburg PA
CBHW030020240426
43672CB00007B/1023